Microwave Engineering - Foundational Studies and Multifarious Applications

Edited by Kim Ho Yeap,
Soo Qi Ping and Veerendra Dakulagi

Published in London, United Kingdom

Microwave Engineering - Foundational Studies and Multifarious Applications
http://dx.doi.org/10.5772/intechopen.1006222
Edited by Kim Ho Yeap, Soo Qi Ping and Veerendra Dakulagi

Contributors
Hirokazu Kobayashi, Jitendra Prajapati, Jyothishree Pillai, Kim Ho Yeap, Lukas Kano Mangalla, Madhur Deo Upadhayay, Tonni Agustiono Kurniawan, Veerendra Dakulagi

First published in London, United Kingdom, 2025 by IntechOpen
IntechOpen is the global imprint of INTECHOPEN LIMITED, registered in England and Wales, registration number: 11086078, 167-169 Great Portland Street, London, W1W 5PF, United Kingdom

For EU product safety concerns: IN TECH d.o.o., Prolaz Marije Krucifikse Kozulić 3, 51000 Rijeka, Croatia, info@intechopen.com or visit our website at intechopen.com.

British Library Cataloguing-in-Publication Data
A catalogue record for this book is available from the British Library

Microwave Engineering - Foundational Studies and Multifarious Applications
Edited by Kim Ho Yeap, Soo Qi Ping and Veerendra Dakulagi
p. cm.
Print ISBN 978-1-83634-182-6
Online ISBN 978-1-83634-181-9
eBook (PDF) ISBN 978-1-83634-183-3

If disposing of this product, please recycle the paper responsibly.

Meet the editors

Kim Ho Yeap holds a Bachelor of Engineering (Honours) in Electrical and Electronic Engineering from Universiti Teknologi Petronas, a Master of Science in Microelectronics from Universiti Kebangsaan Malaysia, and a Ph.D. in Engineering from Universiti Tunku Abdul Rahman. He is currently an Associate Professor at Universiti Tunku Abdul Rahman. A senior member of IEEE, he is also a Chartered Engineer (UK Engineering Council), a Professional Engineer (Board of Engineers Malaysia), a Professional Technologist (Malaysia Board of Technologists), and an ASEAN Chartered Professional Engineer. Yeap has conducted extensive work in electromagnetics. Since 2006, he has secured 3 patents and published 105 refereed journal articles, 51 international conference papers, 15 books, and 28 book chapters. Notably, his published work contributed to the design of receiver optics for ALMA — the world's most powerful millimeter/submillimeter wave radio telescope. He has also served in various administrative roles, including Head of Programme for the Master of Engineering Science, Head of the Department of Electronic Engineering, Chair of the Self-Assessment Committee, and Chair of the Asian Physics Olympiad competition.

Dr. Veerendra Dakulagi holds a Ph.D. in Array Signal Processing from Visvesvaraya Technological University, India, as well as an M. Tech. in Power Electronics and a B. E. in Electronics and Communication Engineering. With over 15 years of experience, he has contributed to more than 60 peer-reviewed publications, including journal articles, conference papers, and book chapters. His groundbreaking work on direction-of-arrival (DOA) estimation and smart antenna systems has earned him multiple international patents. Dr. Veerendra Dakulagi is currently serving as the Professor and Head of the Department of Computer Science and Engineering (Data Science) at Guru Nanak Dev Engineering College (GNDEC), Bidar, India. Dr. Dakulagi is also a Visiting Professor at HELP University, Malaysia (2024–2026). A recipient of numerous accolades, including the INSA Visiting Scientist Award (2023), the IEEE Bangalore Chapter Outstanding Volunteer Award (2021), and the VGST Best Researcher Award (2018), Dr. Dakulagi is recognized globally for his expertise. He has chaired international symposia and actively contributes to IEEE, advancing both research and education. His current research focuses on smart antennas and disaster-response technologies, reflecting his commitment to addressing global challenges through innovation and collaboration.

Soo Qi Ping holds a Ph.D. in Electrical and Electronic Engineering and a First-Class Master's degree from the University of Nottingham Malaysia Campus. He is currently an Assistant Professor at the Faculty of Engineering and Green Technology, Universiti Tunku Abdul Rahman (UTAR), Kampar Campus. Soo is an active member of IEEE and the IEEE Antennas and Propagation Society. He is also a registered Graduate Engineer with the Board of Engineers Malaysia (BEM) and a Graduate Technologist with the Malaysia

Board of Technologists (MBOT). Soo's research expertise lies in electromagnetic wave propagation, particularly in complex environments such as caves, tunnels, and uneven terrains. His contributions have been recognized through multiple accolades, including the IEEE Best Paper Awards in 2018 and 2021. He has served various teaching roles since 2017 and currently lectures in subjects ranging from communication systems to renewable energy. He is also actively involved in academic administration, serving as the Programme Coordinator for the BTech (Hons) in Electronic Systems (Work-Based Learning) and as a member of UTAR's Industrial Training Committee.

Contents

Preface

This book presents a collection of chapters that explore emerging research and applications in microwave and electromagnetic technologies. It begins with an introduction to electromagnetic compatibility (EMC), a critical aspect of modern electronic system design. The chapters that follow delve into key areas, including permittivity measurement using microwave radar, slotted waveguide antenna design for 6G wireless systems, and near-field to far-field transformation techniques for radar imaging. The final chapter demonstrates how microwave radiation can be harnessed for sustainable biomass conversion processes such as torrefaction and gasification. Together, these chapters highlight the versatility and growing importance of electromagnetic theory and microwave technology in various applications, including communication, sensing, measurement, and energy systems. This book is intended for researchers, engineers, and students interested in both the fundamentals and practical applications of electromagnetics.

Kim Ho Yeap and Qi Ping Soo
Universiti Tunku Abdul Rahman,
Perak, Malaysia

Veerendra Dakulagi
Guru Nanak Dev Engineering College,
Bidar, India

Section 1

Fundamentals and Measurement Techniques in Electromagnetics

Chapter 1

Introductory Chapter: An Overview of Electromagnetic Compatibility

Kim Ho Yeap and Veerendra Dakulagi

1. Introduction

The world is now at the cusp of the Fourth Industrial Revolution (IR 4.0). Unlike its three preceding brethren, IR 4.0 involves the extensive application of smart technologies, such as cyber physical systems, the Internet of Things (IoTs), and the Internet of Systems. These technologies aim to improve the quality of life in an exponential manner. The advent of IR 4.0 has summoned a clarion call for the need to accommodate more electronic and/or electrical components within a constrained space and to increase the operating frequencies of these components. Practically speaking, the energy carried by a propagating electromagnetic wave is directly proportional to its oscillating frequency [1]. Hence, when devices operating at high frequency are placed in close proximity, there is a possibility of causing disturbances among themselves. Electromagnetic compatibility (EMC) has therefore become a primary concern to be addressed by engineers working in the electrical and electronic (E&E) industry.

To put it simply, EMC is the ability of an electrical or electronic device to function correctly and accurately, with its performance unaffected by devices in its vicinity and without itself becoming a source of interference to others. This is to say that an electromagnetically compatible device should possess the characteristics of low susceptibility and emissions to its environment [2]. Electromagnetic interference (EMI) usually occurs as a result of signal radiation and conduction. To ensure that the device is immune to EMI, the device itself should be insusceptible to both signals propagating in air and transmission lines. At the same time, it should also minimize the risk of leaking electromagnetic fields and waves to its environment.

2. Common EMI issues

Crosstalk and electrostatic discharge (ESD) are, perhaps, the two most commonly found EMI problems in E&E devices. Crosstalk usually affects both electrical appliances and electronic devices when current in transmission lines experiences noise interference. ESD, on the other hand, affects mostly semiconductor devices. Nevertheless, there are rare cases where ESD causes breakdown in electrical equipment as well. An example of such case is a lightning strike which imposes a sudden voltage spike in the electrical equipment.

Crosstalk occurs when electromagnetic signal, in the form of fringing fields, leaks out from a transmission line and couple to its neighboring line via capacitive or inductive coupling. The noise generated in the transmission line as a consequence of the coupling field causes deterioration to the performance of the E&E devices. Owing to the superposition of noise with the actual input signal, the device may fail to produce the desired output result. In the event a sudden intense glitch is coupled to the transmission line, the device connected to it may be damaged. The risk of crosstalk corresponds proportionately with the increase in frequency of the coupling fields and the reduction of gap distance between the two lines.

Unlike crosstalk which is a contact-free phenomenon, ESD occurs when two oppositely charged materials are in touch, resulting in a sudden temporary flow of electrical current. Static electricity can accumulate in materials through tribocharging or electrostatic induction. Tribocharging is the process of having two materials come in contact for a short period of time before separating. This may cause an imbalance of charge distributions between the two materials. Common examples of tribocharging are passing a plastic comb through dry hair and walking on a carpet. Electrostatic induction happens when a charged object, such as a Van de Graaf generator, is placed in close proximity with a conductive material sequestered from the ground. The charges redistribute within the material, causing an alignment of charges with identical polarity at the surface of the non-grounded material. When the electrostatically charged materials get in touch with another conducting material, momentary current is generated due to charge recombination.

3. Methods to suppress EMI

Some of the methods implemented to suppress EMI are [2, 3]:

i. Shielding. By applying shields, electronic devices could be protected from electromagnetic interference from the exterior; while, at the same time, fields radiation from the device itself shall not disrupt the operation of other devices. The shield could be in the form of a conducting enclosure (refer to **Figure 1**)

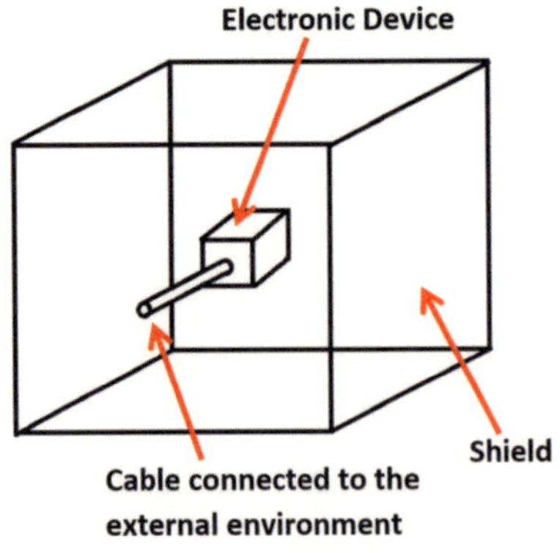

Figure 1.
An electronic device shielded in a metallic box.

which encapsulates the electronic device of parasitic components [4, 5] inserted between transmission lines. It is to be noted, however, that the pigtail of the device, viz., the opening at the end of the transmission line, may expose the device to the risk of crosstalk [2].

ii. Twisting wires. Inductive coupling could be removed when the transmission line is twisted with the ground wire. When crosstalk occurs, Electromotive forces (EMFs) are induced. Due to the opposite polarities in two adjacent loops, however, the EMFs between the loops are canceled. If the number of loops in the twisted wires is even, the net induced EMF could be completely removed. On the contrary, if the number of loops is odd, the net EMF is equivalent to that of a single loop [2]. It is to be noted that if the twisted wires are unbalanced, i.e., the impedances between the wire and the ground are different, only crosstalk contributed by inductive coupling could be eliminated [2].

iii. Balancing. Balancing is a method used to remove common mode noise. The current of a common mode noise flows in the same direction on the positive and negative conductors. The two circuit paths in a balanced circuit share the same ground and both paths comprise identical non-zero impedances [3]. A balanced pair of twisted wires, for instance, has its ground centre tapped. Since the voltage polarities and current directions coupled into the wires oppose each other, the crosstalk noise contributed by both inductive and capacitive couplings can therefore be eliminated [2].

iv. Filtering. To filter common mode and/or differential mode noises, capacitors are added between the positive and negative conductors or inductors are added along the two conductor paths. The current of a differential mode noise flows from the positive to the negative conductors in a single direction.

v. Power supply decoupling. Practical power supplies are a source of noise coupling. To isolate the circuits from their power supplies, resistor-capacitor or resistor-inductor decoupling networks are used [3].

vi. Grounding. There are two types of grounding, i.e., safety and signal grounds. As its name implies, the safety ground is to protect the electrical appliance, the building, and the personnel in the building, when there is a sudden surge of current. The signal ground of a circuit, on the other hand, is the path which allows current to flow back to its source. To suppress the ground noise, the ground impedance and current should be minimized [3].

4. EMC standards

In order to maintain the quality of an E&E product–to ensure that it does not impose EMI risks and that it is insusceptible to EMI disturbance, the product has to conform to certain EMC standards. The standards outline the radiated and conducted limits at different frequencies. If the product is to pass the standards, the detected electromagnetic signal from it has to be less than those stipulated in the standards. In the United States (US), the Federal Communications Commission (FCC) is responsible for

Type	Standard	Area
Commercial	CISPR 11	Industrial, scientific, and medical (ISM) equipment
	CISPR 22	Information technology equipment (ITE)
	IEC 61000–3-2	Harmonics
	IEC 61000–3-3	Flicker
	IEC 61000–4-2	ESD
	IEC 61000–4-3	Radiated immunity
	IEC 61000–4-4	Electrically fast transient (EFT)
	IEC 61000–4-5	Lightning surge
	IEC 61000–4-6	Conducted emission
	IEC 61000–4-8	Magnetic immunity
	IEC 61000–4-11	Dips, interrupts, voltage variations
	IEC 61000–6-1	Immunity for residential, commercial and light-industrial environments
	IEC 61000–6-2	Immunity for industrial environments
	IEC 61000–6-3	Emission standard for residential, commercial and light-industrial environments
	IEC 61000–6-4	Emission standard for industrial environments
	IEC 61000–6-5	Immunity for power station and substation environments
	IEC 61000–6-6	High altitude electromagnetic pulse (HEMP) immunity for indoor equipment
Medical	IEC 60601–1-2	Medical products
Automotive	SAE J1113	Automotive industry
Military	MIL STD 461E	Emission and susceptibility requirements
Aerospace	RTCA DO-160	Emission and susceptibility requirements
	SAE ARP5412B	Aircraft lightning environment and related test waveforms

Table 1.
Common EMC Standards [2, 6, 7].

setting the rules and regulations for E&E devices. The most widely used standard outside the USA is CISPR 22 (for devices operating above 9 kHz) and TC 77 (for devices operating below 9 kHz) published by the International Electrotechnical Commission (IEC). Both FCC and CISPR 22 classify E&E devices into two classes–class A devices are meant for commercial, industrial, and business purposes, whereas class B devices are essentially to be used in the residential environment [2]. In general, the limits set for class B devices are more stringent since the products are mainly used by novices who are neither equipped with the professional knowledge nor expertise to handle the product. Should an unexpected issue arise from it, the possibility of accidents is therefore higher [2]. Some of the popular EMC standards are given in **Table 1**.

5. EMC measurements

To determine if the radiated emission of a device under test (DUT) complies with the EMC standards, tests are usually conducted in an open area test site (OATS) or

Figure 2.
A full anechoic chamber is used for measuring the radiation immunity from a DUT.

an anechoic chamber. An OATS is a wide and flat piece of open land that is free from external electromagnetic signals. It constitutes a layer of highly conducting ground plane which serves as a reflective surface for signal from the DUT to be scattered to the receiving antenna. Unlike the OATS measurement, which is conducted outdoor, measuring the EMC compliance in an anechoic chamber is performed in a room surrounded by ferrite absorber tiles. A semi-anechoic chamber (SAC) is used to measure radiated emission, while a full anechoic chamber (as shown in **Figure 2**) is used to measure radiated immunity. The walls and the ceiling of the anechoic chamber are filled with absorbers. The difference between a full anechoic chamber and an SAC is that the floor of the former is also covered with absorbers while the latter is not [8]. The absorber tiles are used to remove reflections so as to simulate a spacious open area.

6. Conclusion

In this chapter, a brief overview of electromagnetic compatibility (EMC) is illustrated. In general, an E&E device that abides by the EMC standards should possess low radiation and conduct susceptibility and emission. The device should be immune to electromagnetic interference (EMI), introduced particularly by crosstalk and electrostatic discharge (ESD). Some of the methods commonly applied to reduce EMI include shielding, grounding, balancing, filtering, power supply decoupling, and twisting wires. The radiation from a device is typically measured in an open area test site (OATS) or an anechoic chamber to validate its compliance with the EMC standards.

Author details

Kim Ho Yeap[1*] and Veerendra Dakulagi[2]

1 Universiti Tunku Abdul Rahman, Kampar, Perak, Malaysia

2 Guru Nanak Dev Engineering College, Bidar, India

*Address all correspondence to: yeapkh@utar.edu.my

References

[1] Yeap KH, Hirasawa K. Introductory chapter: Electromagnetism. In: Yeap KH, Hirasawa K, editors. Electromagnetic Fields and Waves. London, UK: IntechOpen; 2019. pp. 3-12

[2] Paul CR. Introduction to Electromagnetic Compatibility. US: John Wiley & Sons; 2006

[3] Ott HW. Electromagnetic Compatibility Engineering. US: John Wiley & Sons; 2009

[4] Yeap KH, Cheah ARC, Yeong KC, Hiraguri T, Yokoyama S, Shitara I, et al. Efficiency enhancement in a wireless power transmission system using parasitic components. IETE Journal of Research. 2020;**66**:806-814

[5] Yeap KH. Wireless power transmission: An overview. In: Gregor R, Yeap KH, Nwajana AO, editors. Power Electronics, RF, and Microwave Engineering. London, UK: IntechOpen; 2023

[6] Wyatt K, Jost RJ. Electromagnetic Compatibility EMC Pocket Guide. US: SciTech Publishing; 2013

[7] International Electrotechnical Commission Generic EMC Standards. 2023. Available from: https://www.iec.ch/generic-emc-standards [Accessed: January 1, 2023]

[8] Chung BK. EMI/EMC chamber design, measurement, and instrument. In: Chen ZN, editor. Handbook of Antenna Technologies. Singapore: Springer; 2015. pp. 1-23

Chapter 2

Permittivity Measurement by Imaging Radar

Hirokazu Kobayashi

Abstract

This chapter primarily focuses on characterizing the permittivity of dielectric materials by using microwave imaging radar. The propagation of electromagnetic waves in a medium is governed by a fundamental quantity known as the permittivity. The design of high-frequency devices and the selection of materials depend on the complex permittivity. Various methods have been proposed for measuring this constant, which are employed according to the frequency band and material properties. A microwave image can also be obtained by scanning a mechanical or multiple electrical antennas on a short-range radar. This technique is commonly referred to as synthetic-aperture radar (SAR) or multiple-input, multiple-output (MIMO) radar. In order to compensate for the image-positioning errors caused by microwave phase delay through dielectrics, it is necessary to know the permittivity and thickness of the dielectrics. Therefore, this can be estimated by comparing measurements taken with and without the dielectrics present.

Keywords: permittivity measurement, array-factor focusing, path length, imaging radar, multiple-input multiple-output (MIMO), short-range radar, multilayer dielectrics, wall-through, geometrical theory of diffraction (GTD), physical optics (PO)

1. Introduction

The complex permittivity, also known as the dielectric constant, is an essential and fundamental material property to consider in the analysis and design of electromagnetic wave systems. The electric flux density (**D**) is proportional to the electrical field (**E**), as expressed by $\mathbf{D} = \varepsilon_0 \mathbf{E}$, where the constant ε_0 is defined as the permittivity in a vacuum. On the other hand, the **E** change in a material becomes a complex number because the **D** is delayed in time:

$$\mathbf{D} = \varepsilon \mathbf{E} = \varepsilon_0 \varepsilon_r \mathbf{E} = \varepsilon_0 \left(\varepsilon_r' - j\varepsilon_r'' \right) \mathbf{E}, \quad (1)$$

where they are referred to as such, the real part (ε_r') is the relative permittivity, and the ratio of the real part to the imaginary part ($\varepsilon_r''/\varepsilon_r' = \tan\delta$) is known as the dielectric tangent. This means the real part determines the phase constant (β) of the electromagnetic waves, while the imaginary part determines the attenuation constant (α) involved in absorption and loss. Permeability, or the magnetic material constant μ, can be defined similarly to the permittivity. However, since most materials have nearly the same permeability (μ_0) of a vacuum, the relative permeability (μ_r) is taken to be equal to one in this chapter.

The wavelength of an electromagnetic wave in a material is $1/\sqrt{\varepsilon_r'}$ of that in a vacuum ($\mu_0 = 1$), and the propagation velocity is the speed of light c in a vacuum, whereas in a material it slows down to $c/\sqrt{\varepsilon_r'}$. Electromagnetic waves incident on different materials are partly transmitted and partly reflected, and the transmission and reflection coefficients depend on the electrical constants of the materials. The propagation loss of electromagnetic waves is characterized by the dielectric tangent. Thus, the sign of $j\varepsilon_r''$ is generally negative except in the case of metamaterial characteristics.

Several methods have been proposed for measuring complex permittivity. These methods are selected based on the frequency band and the characteristics of the material being measured. Typical and traditional methods include the transmission-line method, the resonator method, and the free-space method, all of which are detailed in articles [1–4]. The outlines of these methods are given below, including the radar imaging method discussed in this chapter.

a. *Transmission-line method*: When a sample is placed inside a waveguide or coaxial line, the propagation constant and characteristic impedance of the structure change. The electrical constants, such as ε and μ, can then be estimated based on these changes. Waveguides are often used as transmission lines because samples are easy to process, the electrical field direction in the sample is almost constant, and analysis is straightforward. However, if the sample has dimensional errors or there is a gap between the sample and the waveguide wall, ε will be evaluated as small. Additionally, the larger ε, the more pronounced the error tends to be. In a coaxial line, a sample of a known length is placed between the line's inner and outer conductors, and the phase and amplitude of the reflected and transmitted waves change due to the change in transmission impedance. The propagation mode in a coaxial line is an almost TEM-mode at low frequencies. However, if the measurement wavelength is shorter than $\pi(r_a + r_b)$, the TE_{11}-mode is generated. If the sample is tilted or if there is a gap between its contact surfaces and the two conductors, the electric and magnetic fields will not be uniform, causing errors.

b. *Resonator method*: The resonant frequency and the quality value (Q) change when a sample is inserted into positions where the electrical field distribution inside the cavity is either zero or at its maximum. These changes are used to evaluate the permittivity and permeability. The premise is that the perturbation of the resonant frequency is minimal before and after filling the sample, and the principle is to measure at a single frequency. This method is highly accurate. For samples with high permittivity or permeability, the volume is reduced prior to measurement. Inside the resonator, there are regions where only an electric or magnetic field exists, which makes it possible to measure permittivity and permeability independently. It becomes difficult to achieve a high Q with rectangular cavity resonators as the frequency increases, and for samples with large losses, the Q value is low, which affects measurement accuracy. In this case, a circular type is used. Since the cavity resonance method is based on the perturbation method, a small sample size is preferable.

c. *Free-space method*: This method determines the permittivity and permeability of a sample by radiating electromagnetic waves in free space and measuring the reflected or transmitted waves. Typically, the size of the measurement space and the sample are chosen to be several to ten times the wavelength being measured. This method is suitable for wideband measurements and effective in the millimeter wave band, which is difficult to measure using the transmission line or cavity resonator methods. Since the sample is placed on a flat surface, dimensional accuracy is relatively loose, and flat objects are often used. A metal plate with dimensions equal to those of the sample is used as the reference surface for the reflection coefficient. The reflection coefficient of this metal plate is set to $\Gamma = 1$, and the reflection coefficient of the sample is determined by comparing it with that of the metal plate. This method includes reflections from unnecessary parts, so pay close attention to this. Other things to be aware of include isolation between the transmitting and receiving antennas and multipath between the sample and antenna. If the obtained frequency characteristics fluctuate slightly, if the imaginary part of the dielectric constant is negative, or if the magnetic permeability of a nonmagnetic material is far from one, it is often due to the sample deviating from the reference plane.

d. *Image shifting method in this chapter*: The object to be measured is placed between the imaging radar and the reference surface, and the permittivity is calculated by correcting the image position depending on whether or not the object is present. Generally, the prior thickness information is required, and it may be limited to flat objects. It is possible to evaluate the equivalent permittivity of relatively large objects, or objects with material distributed plate-like space. Since it is a method based on image comparison, measurements can be taken with a stable signal-to-noise (S/N) ratio. However, reading errors directly lead to measurement errors. For details of this fourth new method, see the main text.

Now, this chapter discusses the measurement of permittivity using radar images. In general, the accuracy of identifying invisible objects underground or behind walls improves drastically if the target image is high quality, as is the case with the surface-penetrating and the wall-through radar. In the field of imaging processing using microwaves or millimeter waves, the synthetic aperture radar (SAR) or an equivalent method to SAR is employed. These methods involve spatially moving an antenna to obtain a large aperture through signal delay processing [5]. If there is some dielectric between the radar and the object, the position of the radar image will be displaced depending on the electrical and mechanical constants of that dielectric-like object, that is, the equivalent permittivity and the equivalent thickness of the object. From this point, these constants can be inverted by comparing the images with and without the dielectric.

Therefore, we firstly discuss a simple and practical method to obtain a radar image using the aforementioned array-factor (AF) theory in short range, rather than SAR processing, which concerns hardware and software using the azimuth and range compression technologies, and its hardware and software are necessary when full-scale SAR processing is adopted for image processing. Alternatively, we can evaluate the radar image around the target using the well-known antenna engineering theory. The imaging procedure involves comparing the target's coordinates with the phase of the target echo. According to this approach, the radar's hardware and software are straightforward, that is, continuous wave (CW) radar. However, calculating image processing for the wide area is costly. Therefore, this imaging method using array-factor focusing (AFF) may only be applicable to short-range radars.

Secondly, we explain a new measurement method for estimating the equivalent permittivity and thickness of a multilayered dielectric plate using a through-radar approach, taking into account practical applications. When a dielectric object, such as a wall, exists between the target and the radar, the position of the target image shifts along the line between the target and the radar due to electromagnetic wave delay in the dielectric. In order to compensate for this shift error, the equivalent permittivity and thickness of the dielectric are needed. Therefore, we can also calculate the equivalent permittivity and thickness of the wall by comparing it with an image with no objects. However, since the permittivity and thickness of the dielectric object are measured together, a novel approach is required to separate them. Another challenge in compensating for the target image's position error in practical and general applications is that the wall's electrical specifications are usually unknown. However, when the thickness of the sample is known in normal permittivity measurements, there is no separation problem when measuring with the image-comparison method.

2. AFF short-range imaging technology

2.1 Array-factor focusing (AFF)

Radar imaging using AF is the method of correlating the measured phase information with the ideal phase data of the radar echo within the illuminated area. This relationship can be described by a simple exponential series of phases, and the result is considered as a kind of SAR to improve the azimuth resolution. It is not necessary to prepare each element of the array to transmit an electromagnetic wave and receive the echo. If the position of the target and radar are relatively fixed, or assuming the position relationship is maintained in a radar processing time rate, a transmit (Tx) and receive (Rx) antenna can also be scanned mechanically and independently. In addition, it becomes easy to prepare the Tx and Rx antennas separately to avoid the complexity of the microwave circuit under the measurement. For example, when measuring with a general-purpose two-port vector network analyzer (VNA), it is better to use two antennas for Tx and Rx that are spatially separated to maintain isolation between Tx and Rx. And this situation is similar to a MIMO system dealing with multiple transmit and receive signals simultaneously. In this sense, virtual array processing for higher resolution, etc., which is omitted here, may be applicable.

Let us assume that there are multiple point sources with an isotropic spatial pattern for the electromagnetic field formed by these point sources. Considering that each wave source acting as an antenna has the same phase with each other at its feeding point, then its far field can be expressed by the simple superposition of each electromagnetic field, taking into account the phase due to the coordinate difference of each source. We call this AF, which becomes similar to a Fourier series when the source coordinates are arrayed in a flat plane. A realistic antenna has directivity, so this pattern of the array antenna is estimated by weighting the element pattern by the AF. This AF does not include the spatial mutual coupling between the elements, but this fact sometimes results in the occurrence of a difference between measurement and theory, depending on a condition. Under this consideration, we indicate the AF of point sources isolated at any coordinate in space; the angular variable of spherical coordinates is (θ, ϕ), as follows:

$$f(\theta,\phi)=\sum a_n \exp\{jk(x_n u+y_n v+z_n w)\}, \tag{2}$$

where (x_n, y_n, z_n) is the nth element in the three-dimensional (3D) spherical coordinate system ($u = \sin\theta\cos\phi, v = \sin\theta\sin\phi,\ w = \cos\theta$) and a_n is its complex amplitude. Eq. (2) is the electromagnetic field formed in the far region by n isolated point elements. Therefore, if the sampling data coincide with these point sources, we can apply this idea to the transformation of near field to far field in the field of antenna and radar cross-section (RCS) measurement, and we can find that it is possible to evaluate the measurement data even for cylindrical or spherical scanning [6]. Since Eq. (2) is expressed in three dimensions, it is expected that a 3D image can be obtained and that the sampling position of Tx or Rx is independently arbitrary [7]. In this paper, we attempt to further extend and simplify this method.

In order to understand some more details of the imaging process with respect to Eq. (2), consciously taking into account the imaging, we express a model of the scattering field in spatial terms as

$$\exp\{s(x,z)\}=\sum_{m=1}^{M} A_m\cdot\delta(\mathbf{r}-\mathbf{r}_m), \tag{3}$$

where $\mathbf{r}$, $\mathbf{r}_m$ are the position vectors of the observation and the point source (scattering center), respectively, and the Dirac delta function $\delta(\mathbf{r}-\mathbf{r}_m)$ is interpreted as the point scatterer or the pixel position of the image area. From Fourier theory, the $\delta(\mathbf{r}-\mathbf{r}_m)$ is approximately derived as follows [8]:

$$\delta(\mathbf{r}-\mathbf{r}_m)=\int_{-\infty}^{\infty}\exp\{-j(\mathbf{k}\cdot\mathbf{r}_m\}\cdot\exp\{-j(\mathbf{k}\cdot\mathbf{r}\}d\mathbf{k}. \tag{4}$$

In other words, if $E^s(\mathbf{k})$ is a scattering field in the range of the wave number vector $\mathbf{k}$, Eq. (3) is the inverse Fourier transform of that field $E^s(\mathbf{k})$. Therefore, using the Fourier transform of Eq. (3), we get

$$E^s(\mathbf{k})=\sum_{m=1}^{M} A_m\cdot\exp\{-j(\mathbf{k}\cdot\mathbf{r}_m)\}. \tag{5}$$

In Eq. (5), the 2D radar image is given by $\mathbf{r}_m=(x_m,z_m), \mathbf{r}=(x,z), \mathbf{k}=(k_x,k_z)$ in the x-z plane (z: range direction). This equation has a receiving form as radar and is similar to the expression AF, Eq. (2). Therefore, based on Eqs. (2) or (5), the radar image for the coordinates of the target will be obtained according to the focusing processing of the near-region target and the Tx/Rx two-way processing. In addition, this assumption for the target to be plural small spherical points in the above discussion is often used as an approximate method as "point source."

To describe the focusing procedure, we use the rectangular (x,y,z) coordinate system to describe azimuth, elevation, and range from the radar point of view. As shown in **Figure 1**, the coordinates of the m-th Tx antenna are expressed as $\mathbf{r}^t_m=(x^t_m, y^t_m, z^t_m)$, and the coordinates of the n-th Rx antenna are expressed as $\mathbf{r}^r_n=(x^r_n, y^r_n, z^r_n)$. On the other hand, the coordinates of the target are described as $\mathbf{r}_p=\left(x_p, y_p, z_p\right)$ for the image data variable. Here, $\mathbf{r}$ shows the position vector. The transmitted wave from m-th Tx antenna is reflected on the target, which is assumed to be focused at position (x_p, y_p, z_p), and then

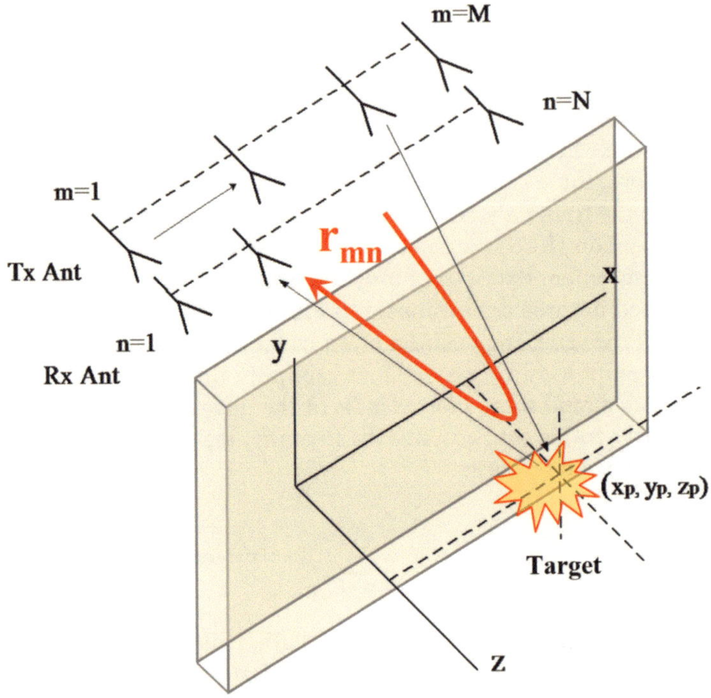

Figure 1.
Radar transmit (Tx) and receive (Rx) antennas for array-factor focusing (AFF) with a wall-like object between the radar and the target.

returns to n-th Rx antenna. At this point, the optical path length from Tx to Rx *via* the target, $r_{mn}\left(x_p, y_p, z_p\right)$, is given as

$$r_{mn}\left(x_p, y_p, z_p\right) = |\mathbf{r}_p - \mathbf{r}^t_m| + |\mathbf{r}^r_n - \mathbf{r}_p|. \tag{6}$$

Since the spatial wavenumber is $k = |\mathrm{k}|$, the phase path length with no object between Tx and Rx is converted to kr_{mn}. Applying this phase change procedure to AF, we can obtain the target image. Eq. (2) is formulated for sources in the infinite far field, whereas Eq. (6) corresponds to "focusing" to the near field or to a similar migration procedure in the SAR field. This suggests that AFF has a higher resolution than SAR for short-range targets.

2.2 Imaging technology using AFF short-range radar

Above, we discussed the spatial imaging procedure for focusing on the near field to obtain a target image at short range. This procedure primarily impacts the azimuth resolution parallel to the antenna array aperture. For the actual measurement, frequency bandwidth is also needed, which affects the range resolution perpendicular to the antenna aperture. The transmit (Tx) waveform is a continuous wave (CW), and the signal is swept in a stepped fashion. Setting the receive power to $P_{\ell mn}\left(f_\ell, r_{mn}\right)$, where m is the Tx antenna, n is the Rx antenna, and ℓ is the step frequency f_ℓ, from the previous discussion, the return signal from the target can be expressed as follows:

$$Q_0(\mathbf{r}) = \sum_{\ell=1}^{L}\sum_{m=1}^{M}\sum_{n=1}^{N} P_{\ell mn}\left(f_{\ell}, r_{mn}\right) \cdot \exp\left\{j2\pi\frac{f_{\ell}}{c} r_{mn}(x,y,z)\right\} \cdot \exp(j\phi_{\ell}), \tag{7}$$

where the target coordinates $\mathbf{r}_p$ are replaced by the imaging area variables $\mathbf{r}(x,y,z)$ and M, N, K are the Tx, Rx antenna position (or antenna element number), and the number of frequency steps, respectively. For the wavelength k, $k_{\ell} = 2\pi f_{\ell}/c$ with the speed of light c. $P_{\ell mn}\left(f_{\ell}, r_{mn}\right)$ is the received power, and then, the complex amplitude a_n in Eq. (2) corresponds to the antenna gain. The receive power is expected to be calibrated by the frequency characteristic of a standard target $\sigma\left(f_{\ell}\right)$. It is also desirable that the phase ϕ_{ℓ}, which is the inner delayed phase quantity of the measuring instrument or radar system, is calibrated by the standard RCS target, such as a conducting sphere. Finally, we can obtain 3D image data by directly substituting Eq. (7) with the variables $\mathbf{r} = (x,y,z)$.

As mentioned above, it is interpreted that the AFF image expresses the relationship between the received signal and the imaging region as the degree of phase matching. This is understood by the AF theory, similar to the AFF imaging concept, that $Q_0(\mathbf{r})$ indicates a peak value in the spatial domain when the difference $\mathbf{r} - \mathbf{r}_p$ of the imaging area variable $\mathbf{r}$ and the target coordinates $\mathbf{r}_p$ has a very close distance. Next, we will improve Eq. (7) by calibrating for image defocus.

The estimation expression $Q_0(\mathbf{r})$ is related to the position of the Tx/Rx array antenna element and the frequency as an exponential periodic function. Therefore, the spatial ambiguity is caused by the state of the parameter. This is a grating lobe occurrence in antenna theory and an alias in Fourier theory, and it is necessary to avoid this nature when designing the AF. In addition, it is better to consider the illumination change due to the antenna's directivity when simulating $Q_0(\mathbf{r})$. As can be guessed from the derivation of $Q_0(\mathbf{r})$, this expression has a Fourier series form of the planar array, so that its resolution in the cross range is approximately half the beamwidth, which depends on the aperture length. There is an unnecessary obstacle between the radar and the target for the sensor, such as a wall-through or underground radar. Assuming that the wall is modeled by multilayer dielectric plates and taking into account the transmit and receive coefficients, we can theoretically express the model.

In this multilayered plate, the wave from the radar propagates to the target with the latest refraction coefficient after multiple reflections and refractions in the wall. On the other hand, as the Tx/Rx point of the antenna array changes with each measurement, the angle of incidence of the wave on the plate is not fixed. At this point, there is a problem, as the wave is delayed in the dielectric material. When the angle reference is facing the target, the wave with a different angle of incidence will always have a delay. It is predicted that this delay will not be so small for close-range radar. In particular, when the aperture length is greater than the size of the target and the angle of incidence is more than a few tens of degrees, the delay can be of the order of centimeters in relation to the distance. In order to confirm this delay, we try to make a simple check using Snell's law.

According to Snell's law, we get $k_1 \sin\theta_1 = k_2 \sin\theta_2$ at the boundary where permittivity and permeability are different in each material. The $k_{1,2}$ is the wavenumber of each material is related to the electric constant, (ε, μ), $k = \omega\sqrt{\varepsilon\mu}$ ($\omega = 2\pi f, f$: frequency). If there are $\mathcal{N}$-layer dielectric plates, then by repeating each plate in turn, the following relationship is obtained as

$$k_1 \sin\theta_1 = k_2 \sin\theta_2 = \cdots = k_i \sin\theta_i = \cdots = k_{\mathcal{N}} \sin\theta_{\mathcal{N}}, \quad i = 1, 2, \cdots, \mathcal{N}. \tag{8}$$

In this expression, which gives the initial angle of incidence θ_1, we can easily calculate the angle of incidence on the i-plate. Now, setting the thickness of each plate to be z_i, using the Pythagorean proposition, the optical path length of the refracted wave through the whole $\mathcal{N}$ plates is given by

$$D = \sum_{i=1}^{\mathcal{N}} k_i z_i \cdot \left(k_i^2 - k_1^2 \sin^2\theta_1\right)^{-1/2}. \tag{9}$$

This expression gives a relationship between the Tx antenna position (x_0, y_0, z_0) and the observation point with final refraction to the target (x, y, z). If two of these three parameters, angle of incidence, Tx antenna, and observation point coordinates, are fixed, the remaining one can be determined automatically.

Taking into account the additional obstacle for the total path length in Eq. (9), Eq. (7) is modified to:

$$\begin{aligned} Q(\mathbf{r}) = \sum_{\ell=1}^{L}\sum_{m=1}^{M}\sum_{n=1}^{N} & P_{\ell mn}(f_\ell, r_{mn}) \cdot \exp\{jkD^t(\mathbf{r}_m, \mathbf{r}, N)\} \\ & \cdot \exp\{jkD^r(\mathbf{r}_n, \mathbf{r}, N)\} \cdot \exp\left\{j2\pi\frac{f_\ell}{c} r_{mn}\right\} \cdot \exp(j\phi_\ell), \end{aligned} \tag{10}$$

where $D^t(\mathbf{r}_m, \mathbf{r}, N)$ and $D^r(\mathbf{r}_n, \mathbf{r}, N)$ are the path length of the N layered wall plate inserted from the Tx and target side, respectively. If $P_{\ell mn}(f_\ell, r_{mn})$ is not measured but calculated for the simulation; this expression becomes

$$\begin{aligned} P_{\ell mn}(f_\ell, r_{mn}) = {} & G^t_{\ell mn}(f_\ell, r_{mn}) \cdot T^t_{\ell mn}(f_\ell, r_{mn}) \\ & \cdot \sigma(f_\ell, \theta_{in}, \varphi_{in}; \theta_{out}, \varphi_{out}, pol) \cdot G^r_{\ell mn}(f_\ell, r_{mn}) \cdot T^r_{\ell mn}(f_\ell, r_{mn}), \end{aligned} \tag{11}$$

where $T^t_{\ell mn}(f_\ell, r_{mn})$ and $T^r_{\ell mn}(f_\ell, r_{mn})$ are the refraction coefficients of the $\mathcal{N}$-layered plate from the Tx and Rx sides, respectively, which is reviewed in Appendix A, $G^t_{\ell mn}(f_\ell, r_{mn})$ and $G^r_{\ell mn}(f_\ell, r_{mn})$ are Tx and Rx antenna (gain) patterns, respectively, and $\sigma(f_\ell, \theta_{in}, \varphi_{in}; \theta_{out}, \varphi_{out}, pol)$ is the RCS of the target with incident angle, scattering angle, and field polarization denoted by "pol." For the σ it is better to consider its near field as described in Refs. [9, 10]. As the antenna pattern affects the weighting of the incident and scattered waves from the target, this compensation cannot be ignored, especially in the near field region.

Finally, in this section, we show the AFF wide area measurement image example in **Figure 2** as a result. The image size is $7 \times 10\ \text{m}^2$ with the origin at the center of the antenna aperture, which is vertically polarized, all subsequent measurements are taken with vertical polarization. The lower and upper maps show the 2D and 3D images, respectively, with pixel strength on the vertical axis. The target is a metal strip. There is also a measurement pedestal and absorption material for passage use behind the target, but it is not a wall case. As can be seen in the figures, there is a locus with a relatively large reflection level, which corresponds to the rectangular frame of the radio anechoic chamber. It is noted that there is all the information is in the received reflection data as the waveform in the time domain is CW mode. Therefore, the superposition of the data to AF is interpreted as extending the information stored

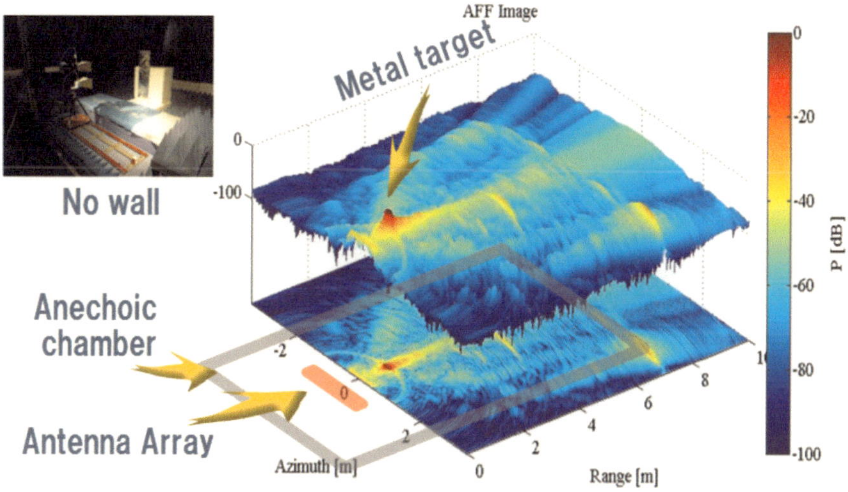

Figure 2.
Measured wide area image around anechoic chamber (7×10 m).

in the received CW signal to the spatial domain. Originally, AF has the same expression as the Fourier series, so it is natural to consider the AF image itself as the spatial spectrum of the received signal.

Now, we consider the geometrical theory of diffraction (GTD) near field modeling and compare it to the measurement. GTD is a well-established high-frequency calculation method for electrically large objects, which extends the geometrical optics (GO) to the concept of diffraction rays using Fermat's principle [10, 11]. In this section, we use the uniform asymptotic theory (UAT) method, which has obtained expressions of diffraction from the edge condition without comparison with the canonical solution, like GTD. Thus, all GTD/UAT use the ray concept, whose main applications are limited to uncomplicated shaping objects or 2D problems, but gives an accurate field solution for near field.

The conducting strip is considered a combination of two conductive half-planes. There are then two edges in the 2D problem. The total field consists of four waves, namely a reflected wave from the strip, two diffracted waves from each of the two edges, and an incident wave from a line source. The two-face corner reflector consists of two strips, and there are five waves to the observation point, namely two single or double reflected waves from the strip, two diffracted waves from each edge, and an incident wave from a line source, as shown in **Figure 3**. For the two-face reflector, the existence of a double- or single-reflected wave depends on whether the reflection point is inside or outside the reflecting shadow boundary. The RCS patterns of the two-face corner reflector are shown in **Figure 4**, where the size is $5\lambda \times 5\lambda$, the observation point and the line source position are $\rho = 15\lambda$ and $d = 10\lambda$, respectively. These results are checked against exact solutions and are in close agreement. Similarly, we can formulate a 3D corner reflector by combining half-planes. The GTD/UAT modeling is available for near field scattering targets (see Appendix B).

The calculated result of UAT is shown in **Figure 5**, which is the near field model of a conducting strip with E-polarization, compared with the conventional physical optics (PO) as the far field model [10]. The measurement frequency is C-band, and the bandwidth is 34%. The plate is $R = 1$m (1.4λ) away from the antenna. The near field UAT is found to be more accurate than PO. The comparison between calculation

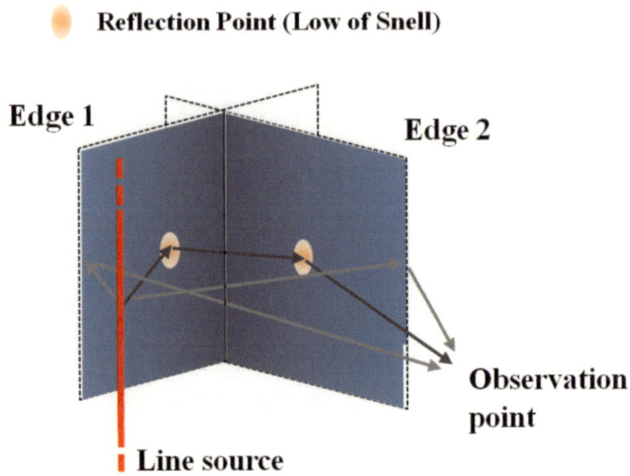

Figure 3.
Two-face corner reflector geometry and its images for GTD/UAT.

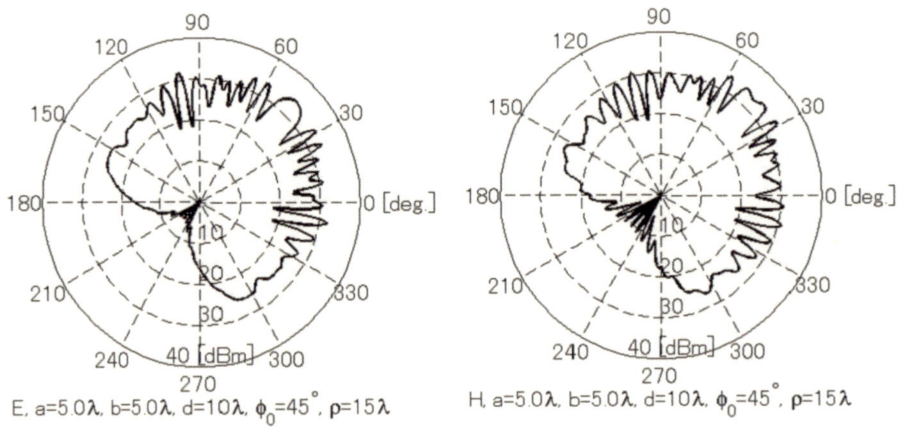

Figure 4.
Calculated near field RCS patterns of two-face reflector by UAT at incident $\phi_o = 45^\circ$, left: E-pol., right: H-pol.

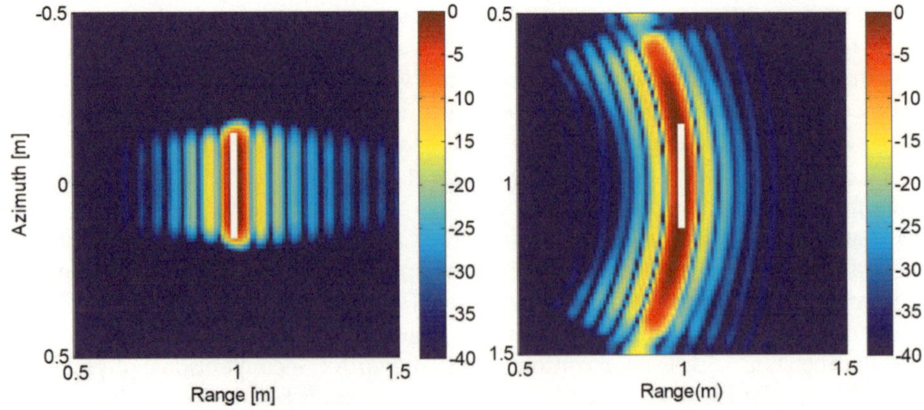

Figure 5.
Calculated images of conducting strip, left: GTD/UAT, right: PO. White line inserted later.

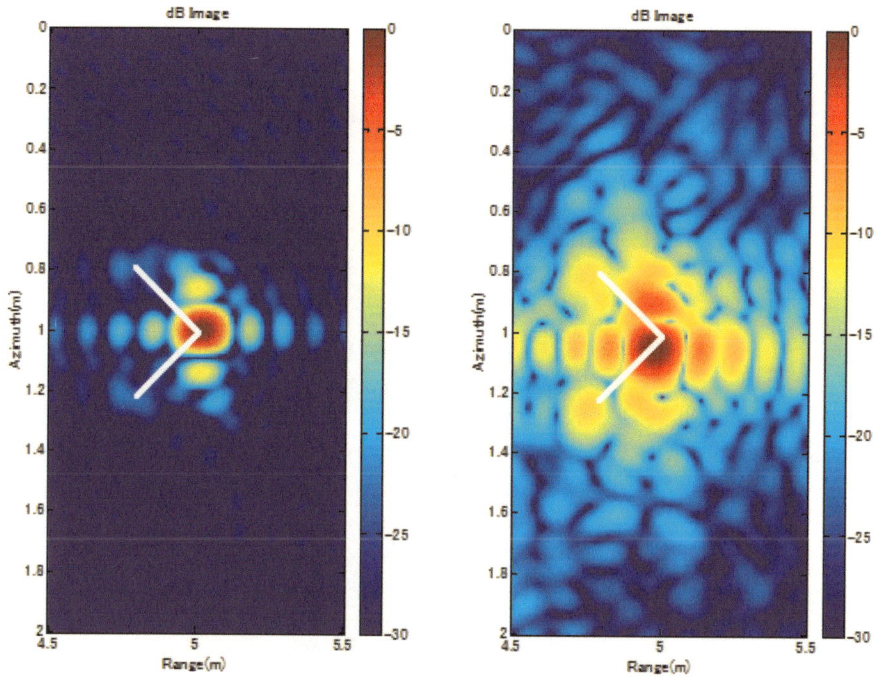

Figure 6.
Two-face reflector images, left: calculated by GTD/UAT, right: measured image. White line inserted later.

and measurement is shown in **Figure 6**, which is the case of a two-face corner reflector. Both figures show the edge diffracted waves, which are interesting to appear to radiate from the edge. The actual targets are also shown in white in these figures. In addition, compared with the traditional SAR image using measured data, the SAR image in the near-region does not show good results. The so-called SAR technology is basically a mono-static mode, while AF radar in measurement is a multi-static mode to be extended to MIMO radar.

3. Estimation of permittivity and thickness using AFF image

First, we discuss the measurement result for a single-layer plate, which is a concrete material with a thickness of 6cm. The distance between the antenna and the target is 100 cm, and the length of the array aperture and the width of the metal plate target are 90 and 22 cm respectively. The measuring Tx center frequency is 4.1 GHz with a bandwidth of 48%. The AFF images are shown in **Figure 7** when the relative permittivity of the concrete is set to $\varepsilon_r = 5.4$. **Figure 7(a)** is no wall, **(b)** is with the concrete wall before compensation, and **(c)** is after compensation. The figure is data without the shifting procedure for the system delay $e^{j\phi_e}$, so that the center of the target appears at 127 cm in the range direction. It can be seen that **(c)**, which is compensated by using Eq. (11), is somewhat improved compared to **(b)**. If there is no information on the specific permittivity, we can obtain it to compare the target position of no wall with that of a wall and convert the different distances into permittivity. However, for general application, it is not able to compensate for image position because we have no information about the wall permittivity, thickness, and

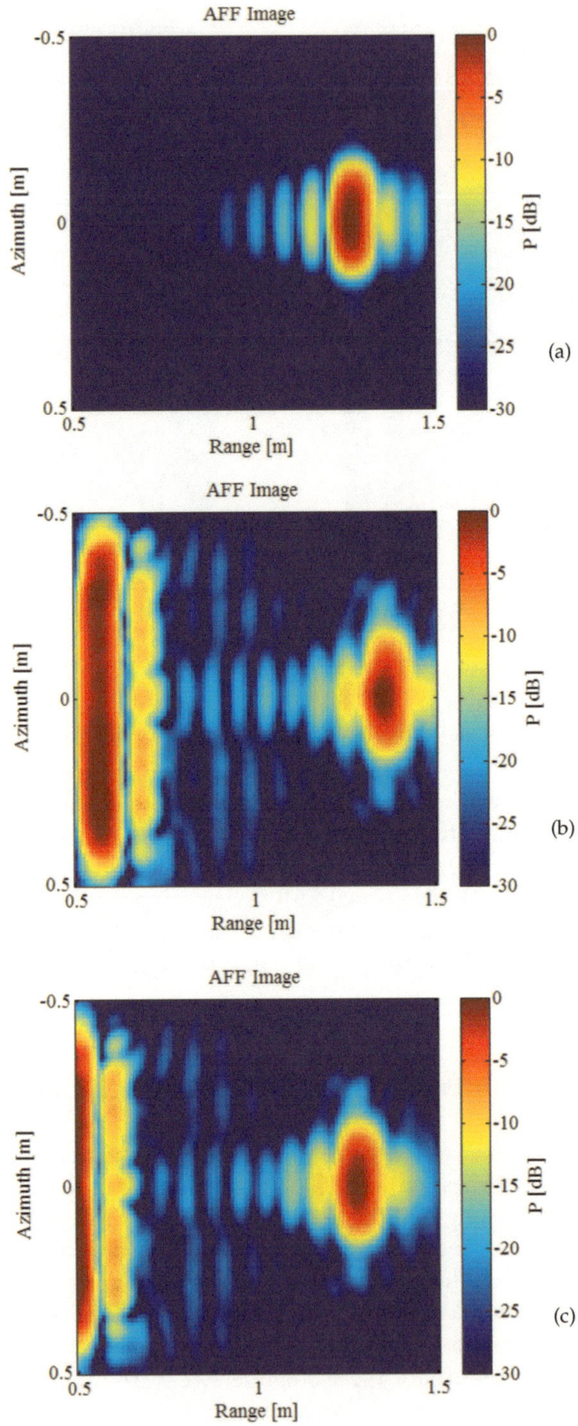

Figure 7.
Measured AFF images of one-layered wall with/without path-length compensation, (a) no dielectric plate (wall), (b) no compensation for position, (c) after compensation.

position. We will discuss later this novel measurement method, the image shifting method (ISM). From this point forward, we will only discuss the dielectric constant in terms of its real part.

It is also capable of measuring the equivalent permittivity of a three-layer wall as well as a one-layer dielectric wall using ISM. Whether the thickness and permittivity are known or unknown, the two-time measurement with and without a wall is an alternative, if the positional shift of the image obtained can be compensated. In other words, it is worth noting that the measurement of (equivalent) permittivity is itself a measurement method. Generally speaking, there are typical measurement methods of permittivity, such as coaxial cable, waveguide, and cavity, and reflection coefficient in spatial wave, as previously mentioned. The former, using a transmission line, is available to obtain an accurate measurement, but is not available for wide area scattering material. The latter method is often used to measure radio-absorbing material, but is unstable to maintain signal-to-noise ratio. The ISM using AFF described in this chapter has a high S/N ratio due to high-gain array antenna image and is applicable to measure the permittivity of scattering material in wide areas, such as bush, tree, and forest, by attaching the metal reference plate.

Thus, the two-time measurements by "with dielectric wall and without" are alternatives in known or unknown of the permittivity and the thickness of the dielectric wall. The "with wall or not" is nothing but the dielectric is different before and after the measurement. Assuming that it is possible to measure twice, the ISM method is applied directly to wall-through radar. The permittivity and thickness of a wall are usually unknown. Therefore, we first measure the wall and then measure this wall again with a dielectric plate whose permittivity and thickness are known, which is prepared beforehand. We can then estimate the equivalent permittivity and thickness of the unknown object by compensating (ISM) the image obtained. At this stage, it is necessary to use the theory of reflection and refraction for multilayer dielectrics. This exact analysis is described in Appendix A. On the other hand, there is also the approximate method, which will be discussed later.

We begin by formulating the ISM in terms of AFF. The velocity of the wave is given as $v = 1/\sqrt{\varepsilon\mu}$ in the dielectric material, where ε and μ are the permittivity and permeability, respectively. If the relative permittivity and permeability are introduced as $\varepsilon_r = \varepsilon/\varepsilon_0$ and $\mu_r = \mu/\mu_0$, respectively, then the wave velocity in the dielectric material is expressed as

$$v = 1/\sqrt{\varepsilon\mu} = c/\sqrt{\varepsilon_r}, \tag{12}$$

where assuming nonmagnetic material $\mu_r = 1$, and $c = 1/\sqrt{\varepsilon_0\mu_0}$ is the speed of light in vacuum or free space (air). In this case, the wavenumber in the material is given by $k = \sqrt{\varepsilon_r}\omega/c = N_{ind}\omega/c$, where N_{ind} is the refractive index $N_{ind} = \sqrt{\varepsilon_r}$ looking from the vacuum or air side. Using the above denominations, the wave toward $+z$ can be expressed by the electromagnetic transmission line model as follows:

$$E_z = E_0 \cdot \exp\{-\alpha z\} \cdot \exp\{j(\omega t - \beta z)\}, \quad jk = \alpha + j\beta, \tag{13}$$

where α and β are the attenuation and phase constants, respectively. If the medium is a lossless material, then it becomes $k = \beta$.

Obviously, from Eq. (13), the path length of the wave in the medium as an insertion phase is given by

$$\Delta\phi = k_0\sqrt{\varepsilon_r}\ z_0 - k_0 z_0 = k_0(\sqrt{\varepsilon_r} - 1)z_0, \tag{14}$$

where z_0 is the thickness of the dielectric object. This phase produces the image shift value $\Delta\phi/k_0$. When converting to the inserted phase, a one-way extra phase $\Delta\phi = k_0 L = k_0(\sqrt{\varepsilon_r} - 1)z_0$ is added for the position shift L on the image. From this consideration, a very simple estimation formula can be obtained as follows:

$$\varepsilon_r = \left(\frac{L}{z_0} + 1\right)^2 \geq 1. \tag{15}$$

If we read the image displacement length from the obtained image, called the reading error $\pm\Delta L$, the estimated permittivity changes with respect to the first order of ΔL as follows:

$$\varepsilon_r \ \rightarrow \ \varepsilon_r \pm \sqrt{\varepsilon_r} \cdot \Delta L / z_0. \tag{16}$$

The error decreases for larger z_0 and increases relatively for larger $\sqrt{\varepsilon_r}$. When the insertion loss of the dielectric plate is IL, the attenuation constant α can be calculated as $\alpha = \ln(IL)/z_0$, allowing the complex permittivity to be evaluated.

Estimating the measurement case of **Figure 7**, since the thickness of the concrete is $z_0 = 6.0\,\text{cm}$ and the positional shift on the image is $L = 8.0\,\text{cm}$, we can calculate the permittivity $\varepsilon_r = 5.4$ using Eq. (14). If the reading error of the image position is assumed to be $\pm 0.5\,\text{cm}$, the estimation error of the permittivity is approximately $\Delta\varepsilon_r = \pm 0.2$. Additionally, the permittivity was estimated to be 5.1 using the coaxial TEM-mode in the transmission line method, and it is described as $\varepsilon_r = 4 \sim 10$ in the Ref. [12]. The ISM measurement method has the advantage of maintaining a high signal-to-noise ratio due to the image compared to a single-beam radar. Furthermore, if it is necessary to obtain the imaginary part of the permittivity, in other words, the loss of the material, it is available by measuring the power with or without the material. In the case of **Figure 7**, we obtain $\varepsilon_r = 5.4 - j\ 0.29$ for the concrete board.

As a second measurement case, we show a three-layer slab consisting of the same concrete as 6 cm thickness and two plasterboards of 9.6 mm thickness on both sides of the concrete. The AFF images in this case and their cross section are shown in **Figures 8** and **9**, respectively. The permittivity of the plasterboard is unknown, while the permittivity and thickness of the concrete are known to be $\varepsilon_r = 5.4$ and 60 mm. Using this information, we calculated the permittivity of the plasterboard to be $\varepsilon_r = 4.0$. The permittivity of this board is described as around $\varepsilon_r = 5$ depending on the material composition, in the general books.

From the above discussion, we can equivalently obtain a set of permittivity and thickness of the unknown wall. For a plate material whose information is unknown, that is, whose information can be evaluated by attaching another plate with known information to this unknown plate. The main problem is the separation of thickness and permittivity. We will briefly discuss this topic.

Now, a wall consists of N layered plates without both side air layers. At this point, although the reflection and refraction waves will move with complicated behavior at the layer boundary, the inserted phase of each layer is approximately given by $k_0\sqrt{\varepsilon_i}z_i$, where ε_i and z_i are the relative permittivity and thickness of the i-th layer, respectively.

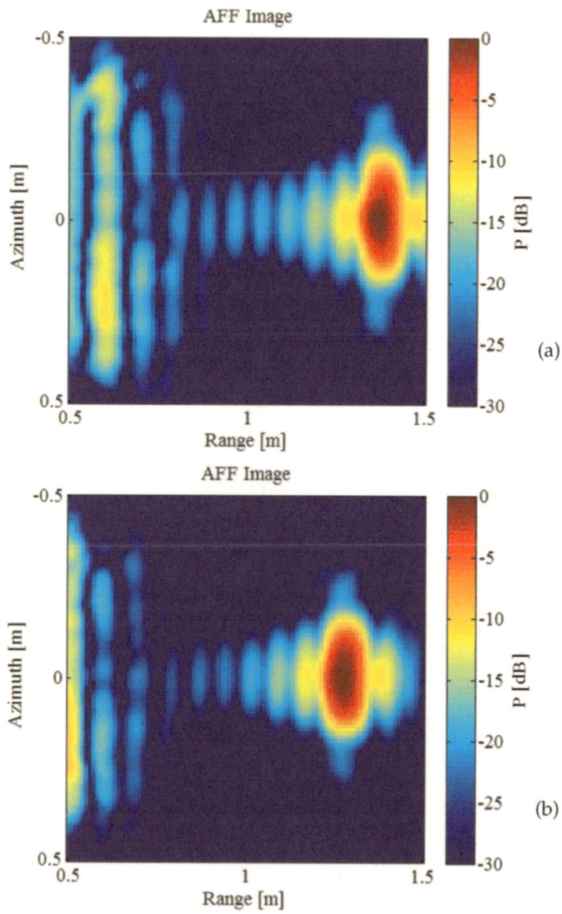

Figure 8.
Measured images of a three-layered wall with/without compensation, (a) before compensation, (b) after compensation.

Taking into account also the angle of incidence θ_0 measured from the normal plane, the inserted phase is modified to $k_0\sqrt{\varepsilon_i}z_i \sec\theta_0$. Therefore, by simply adding the inserted phase of each layer, its relation is derived similarly to Eq. (14) as follows:

$$\sum_{i=1}^{N}(\sqrt{\varepsilon_i}-1)z_i \sec\theta_0 = L. \tag{17}$$

By attaching a few numbers of other dielectric plates with known information, we expect to be able to measure the equivalent permittivity of multilayer dielectrics without comparing no-wall data. From Eq. (17), we can derive the equivalent permittivity of multilayer dielectric plates as

$$\sqrt{\varepsilon_{eq}} = \sum_{i=1}^{N}(z_i\sqrt{\varepsilon_i}) \Big/ \sum_{i=1}^{N} z_i. \tag{18}$$

For example, if $N=2$ and $\theta_0=0$, we get $\sqrt{\varepsilon_2}z_2 - z_2 = -\left(\sqrt{\varepsilon_1}z_1 - z_1\right) + L$. If the right-hand side is a known quantity, the unknown properties of the dielectric can be

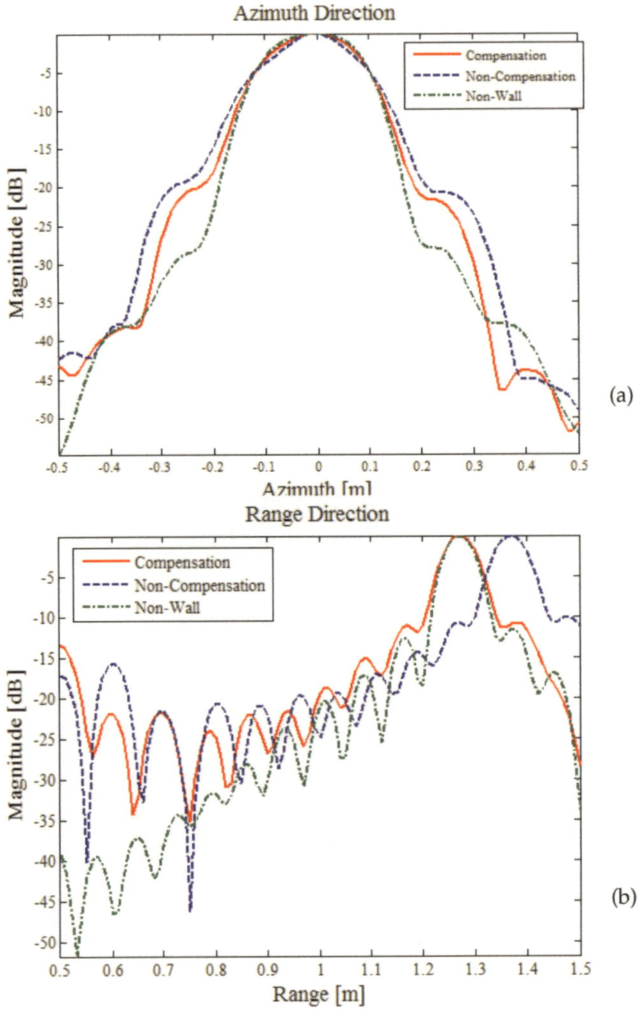

Figure 9.
Magnitude in cross-section of ***Figure 8****, (a) cross-profile in azimuth direction, (b) range profile in z-direction.*

determined. In other words, the unknown properties can be inferred by attaching a dielectric with known properties to one with unknown properties. Unfortunately, it is not possible to separate thickness and permittivity by changing the angle of incidence θ_0, because the extension of z_0 is only projected to L.

For this separation problem of permittivity ε and thickness z, it has been creative so that the attaching plates with different characteristics take two-time measurements. We assume the permittivity and thickness of the first attached plate are ε_a and z_a, and the second plates are ε_b and z_b, respectively. These quantities are known in advance. Then, assuming that the equivalent permittivity estimated by the first and second measurements using ISM is $\varepsilon_{eq,a}$ and $\varepsilon_{eq,b}$, respectively, the desired thickness z_x of the plate under test with unknown electrical constants is obtained from Eq. (18) as follows:

$$z_x = \frac{\left(z_a\sqrt{\varepsilon_a} - z_b\sqrt{\varepsilon_b}\right) - \left(z_a\sqrt{\varepsilon_{eq,a}} - z_b\sqrt{\varepsilon_{eq,b}}\right)}{\sqrt{\varepsilon_{eq,a}} - \sqrt{\varepsilon_{eq,b}}}. \tag{19}$$

The desired permittivity ε_x is derived from Eq. (15) using the above z_x. This expression is the new result through the two-step measurement method.

4. Conclusions

This chapter has discussed methods for evaluating the permittivity of electrical materials. Permittivity is a fundamental constant in science and engineering. Different materials have different permittivities. Many methods for evaluating permittivity are already known in the literature; however, we have shown that it can also be measured with radar images in this chapter. Therefore, we first demonstrated that the AFF imaging method is a useful version of AF with focusing technology. AFF is a simple, yet practical, measurement technique for permittivity. The validity and usefulness of AFF are confirmed by theoretical simulations, high-frequency scattering, and experimental measurement images.

With the AFF method, the equivalent permittivity of a dielectric material with unknown electrical properties can be estimated. This method compares the position of the image of the dielectric under test with that of a non-dielectric. Therefore, this method is highly flexible for relatively large objects distributed in space, such as multilayer plates and forests, which are difficult to measure. ISM also has a high signal-to-noise ratio, making it much more stable than the transmission line, namely waveguide-filling method.

To estimate the unknown thickness and permittivity of the dielectric plate simultaneously, the two-step measurement method is an approximation, but it is easy and practical. Alternatively, one can judge based on the range profile in the obtained image or on reflection and refraction theory (see Appendix A). Higher resolutions through MIMO processing with virtual arrays are recommended for this purpose.

Acknowledgements

This work was partly carried out at Niigata University in Japan. We would like to thank Dr. Yoshio Yamaguchi, Professor Emeritus at Niigata University, for his collaboration.

Appendix A. Reflection/refraction of multilayer dielectric plates

In this appendix, as shown in **Figure 10**, it is shown that the multiple reflection and refraction coefficients for multilayer dielectric flat plates taken from Ref. [10]. Originally, it has been analyzed by Born [13]. For scattering by a target having dielectric parts, the introduction of the surface impedance boundary condition makes it easy to formulate the electromagnetic relations. On the other hand, we can also obtain an exact general formulation for this model, such as multilayer dielectric plates, which are infinitely flat. As a result, the reflection and refraction coefficients obtained from the wave equation are exactly formulated by the characteristic matrix of dielectric permittivity, magnetic permeability, and thickness of each dielectric plate. Furthermore, we can also calculate the coefficient of a metal-backed plate by letting the permittivity of a plate be infinite and the dielectric loss by considering the complex permittivity.

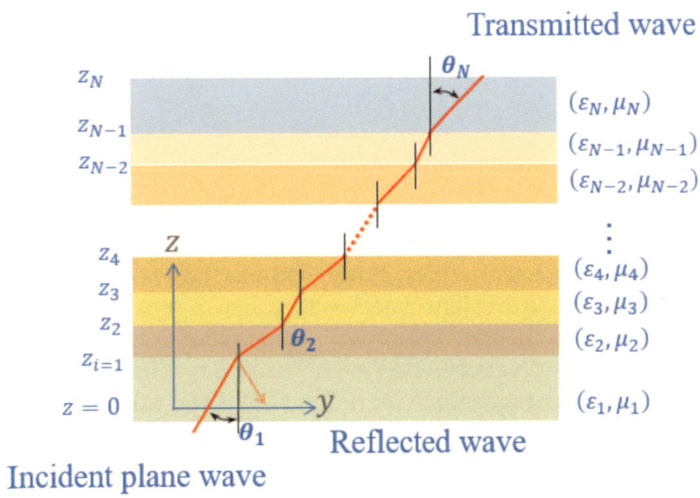

Figure 10.
Multiple reflection and refraction between plural dielectric flat plates.

As shown in **Figure 10**, N-layered plates of infinite length along the y-axis are overlapped along the z-axis. The angle of incidence is θ_1 measured from the normal z-axis of a first plate, and ε_i, μ_i $(i = 1 \sim N)$ are permittivity and permeability, respectively, which are functions of the variable z. In the case of a TE-wave, the electrical field vector is perpendicular to the plane of incidence, the components y, z of the electrical field are 0, and then only a group of (E_x, H_y, H_z) remains. Concentrating on the electrical field component E_x, Maxwell's equation is reduced to three equations:

$$\frac{\partial E_x}{\partial z} + j\omega\mu H_y = 0, \quad \frac{\partial E_x}{\partial y} - j\omega\mu H_z = 0, \quad \frac{\partial H_z}{\partial y} - \frac{\partial H_y}{\partial z} - j\omega\varepsilon E_x = 0. \tag{20}$$

Assuming permeability depends on $\mu = \mu(z)$ and noting $\frac{\partial}{\partial z}(\mu H_y) = H_y \frac{d\mu}{dz} + \mu \frac{\partial H_y}{\partial z}$, we can obtain the following basic equation:

$$\left\{ \frac{\partial^2}{\partial y^2} + \frac{\partial^2}{\partial z^2} + n^2 k_0^2 \right\} E_x = -j\omega H_y \frac{d\mu}{dz} = \frac{\partial(\log \mu)}{\partial z} \frac{\partial E_x}{\partial z}, \tag{21}$$

where $\omega^2 \varepsilon\mu = (nk_0)^2$. After some calculation using the variable separation method, the following characteristic matrix for a dielectric plate is obtained as follows:

$$M(\xi, \varphi) = \begin{pmatrix} \cos\varphi & \frac{j}{\xi}\sin\varphi j\xi\sin\varphi & \cos\varphi \end{pmatrix}, \quad \varphi = nk_0 z\cos\theta, \quad \xi = \begin{cases} \sqrt{\varepsilon_r/\mu_r}\cos\theta : \text{TE wave} \\ \sqrt{\mu_r/\varepsilon_r}\cos\theta : \text{TM wave.} \end{cases} \tag{22}$$

Letting the thickness of each plate is $z_i, i = 1 \sim \mathcal{N}$, each characteristic matrix is recurrently obtained as $i = 1 : G_0 = M_1 G(z_1), i = 2 : G(z_1) = M_2 G(z_2)$, $i = 3 : G(z_2) = M_3 G(z_3), \cdots, i = \mathcal{N} : G(z_{\mathcal{N}-1}) = M_{\mathcal{N}} G(z_{\mathcal{N}})$. Therefore, the matrix for a set of the dielectric plates $i = 1 \sim \mathcal{N}$ is given by

$$G_0 = M_1 G(z_1) = M_1 M_2 G(z_2) = \cdots = M_1 M_2 \ldots M_{\mathcal{N}} G(z_{\mathcal{N}}) = M(z_{\mathcal{N}}) G(z_{\mathcal{N}}). \quad (23)$$

The relationship between matrix elements and coefficients can be derived as follows. The angles of incidence at the first ($i = 1$) and last boundary ($i = \mathcal{N}$) are θ_1 and $\theta_{\mathcal{N}}$ respectively. The amplitudes of the incident, reflected, and refracted waves are A_i, A_r, A_t, the reflection and refraction coefficients are defined as $R = A_r/A_i$ and $T = A_t/A_i$, respectively. Then let m be four elements of the matrix M and consider the constraint conditions $A_i + A_r = (m_{11} + m_{12}\xi_{\mathcal{N}})A_t$ and $\xi_1(A_i - A_r) = (m_{21} + m_{22}\xi_{\mathcal{N}})A_t$, where the coefficients are finally obtained as

$$R = \frac{A_r}{A_i} = \frac{m_{11}\xi_1 - m_{12}\xi_1\xi_{\mathcal{N}} - m_{21} - m_{22}\xi_{\mathcal{N}}}{m_{11}\xi_1 - m_{12}\xi_1\xi_{\mathcal{N}} + m_{21} + m_{22}\xi_{\mathcal{N}}}, T = \frac{A_t}{A_i} = \frac{2\xi_1}{m_{11}\xi_1 + m_{12}\xi_1\xi_{\mathcal{N}} + m_{21} + m_{22}\xi_{\mathcal{N}}}. \quad (24)$$

For example $\mathcal{N} = 3$, let first and third plates are free space, and the permittivity and permeability of the second ($i = 2$) plate are ε_2 and μ_2, respectively. For this case, the characteristic matrix is calculated as

$$m_{11} = m_{22} = \cos\varphi, \quad m_{12} = \frac{j}{\xi_2}\sin\varphi, \quad m_{21} = j\xi_2 \sin\varphi, \quad \varphi = n_2 k_0 t \cos\theta_2, \quad \xi_2 = n_2 \cos\theta_2. \quad (25)$$

Substituting $\mathcal{N} = 3$ into the above equations gives the coefficients as:

$$R = \frac{R_{12} + R_{23}\exp(-j2\varphi)}{1 + R_{12} + R_{23}\exp(-j2\varphi)}, \qquad T = \frac{T_{12}T_{23}\exp(-j\varphi)}{1 + T_{12}T_{23}\exp(-j2\varphi)}, \quad (26)$$

where TE-wave: $n_{1,2} = \sqrt{\varepsilon_{r(1),r(2)}/\mu_{r(1),r(2)}}$ and

$$R_{12} = \frac{\xi_1 - \xi_2}{\xi_1 + \xi_2} = \frac{n_1\cos\theta_1 - n_2\cos\theta_2}{n_1\cos\theta_1 + n_2\cos\theta_2}, R_{23} = \frac{\xi_2 - \xi_3}{\xi_2 + \xi_3} = \frac{n_2\cos\theta_2 - n_3\cos\theta_3}{n_2\cos\theta_2 + n_3\cos\theta_3}, \quad (27)$$

$$T_{12} = \frac{2\xi_1}{\xi_1 + \xi_2} = \frac{2n_1\cos\theta_1}{n_1\cos\theta_1 + n_2\cos\theta_2}, \quad T_{23} = \frac{2\xi_2}{\xi_2 + \xi_3} = \frac{2n_2\cos\theta_2}{n_2\cos\theta_2 + n_3\cos\theta_3}. \quad (28)$$

As expected, these coefficients are available for a single dielectric slab and agree perfectly with Fresnel's slab formula. These formulations are an exact solution of Maxwell's equations and can be applied to dielectric flat plates with losses.

The above theoretical values are then compared with the measured values of a three-layer sandwich radome consisting of skin layers on both sides and a low-density honeycomb core layer. This type has a broadband frequency characteristic and is relatively light. The relative permittivity and thickness of the skin section are $\varepsilon_r = 4.3$ and $t = 0.6$ mm, respectively, while the core section has $\varepsilon_r = 1.1$ and $t = 3.0$ mm. The size of the prototype three-layer radome is 1.2×1.2 in m^2 and is nonplanar with a radius of curvature of 1 m, which is shown in **Figure 11**. **Figure 12** shows the theoretical and measured results for normal incidence, which are in good agreement. For details of mathematical analysis, refer to Ref. [14].

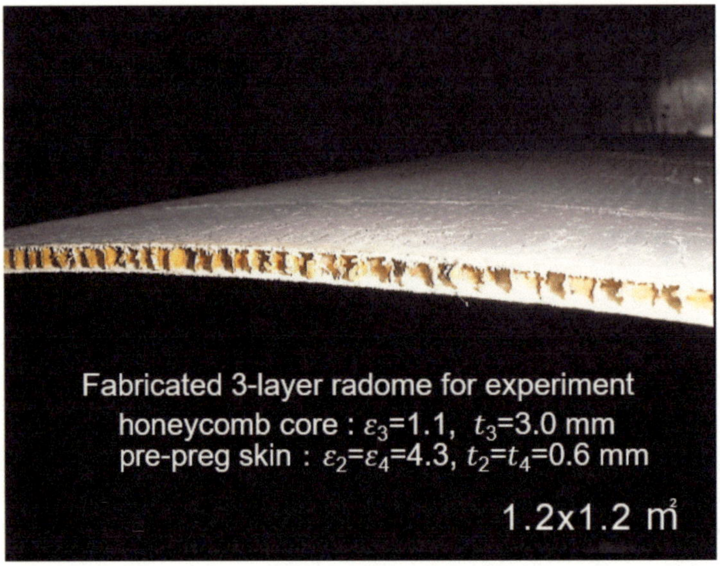

Figure 11.
Three-layer sandwich radome consisting of two high-density skin layers on either side of a low-density honeycomb core.

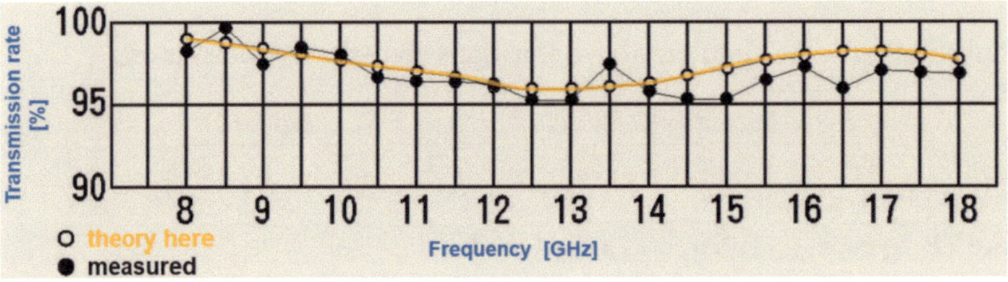

Figure 12.
Calculated and measured power transmission rate characteristics in % and frequency range 8–18 GHz.

Appendix B. Geometrical Theory of Diffraction (GTD) as target modeling tool

Appendix B provides an overview of GTD (GTD), a traditional, high-frequency, asymptotic diffraction theory. The GTD or uniform theory of diffraction (UTD) is a well-established high-frequency prediction method proposed by Keller in '50 for electrically large objects, which extends geometrical optics (GO) to the diffraction ray concept using Fermat's principle [11]. The original GTD is formulated from the exact solution of half-plane diffraction, so the required coefficients are simply expressed by elementary functions and become infinite at the shadow boundary (SB). UTD is an improved theory that avoids this defect of GTD by using Fresnel's integral instead of an elementary function. In this appendix, we use the uniform asymptotic theory (UAT) method, which was mainly established by S. W. Lee in '70 and has obtained expressions of diffraction from edge conditions without comparison with the canonical solution. All GTD/UAT/UTD use the concept of ray; therefore, the main applications are limited to uncomplicated shaping objects or 2D problems.

This appendix describes the near field scattering of a conducting strip and a two-face corner reflector when a local source is radiating. If the incident wave is a plane wave, its scattered far field is easily obtained as a "sinc" function by the method of physical optics (PO) [15]. However, as mentioned above, it is simpler and straightforward for this problem to use the ray technique, such as UAT, since we concentrate the scattering near field from the target.

The conductive strip is considered a combined conductive two half plane. Then there are two edges in the 2D problem. The total field consists of four waves, namely a reflected wave from the strip, two diffracted waves from each of the two edges, and an incident wave from a line source, as shown in **Figure 13**. The strip is placed in the x-y plane. Angular values from the origin, edges, and distances are given in the figure. The two-sided corner reflector consists of two strips, and there are five waves to the observation point, that is, double or single reflected waves from the strip, two diffracted waves from each edge, and an incident wave from a line source as shown in **Figure 13**.

First, the scattered field through a half-plane is expressed as

$$u_{hp}^{t}(\mathbf{r}) = u^{G}(\mathbf{r}) + u^{d}(\mathbf{r}) \tag{29}$$

where u^G is modified geometrical term given by

$$u^{G}(r) = \left[\frac{F(\xi^{i}) - \exp\{-j(\pi/4 + \xi^{i2})\}}{2\sqrt{\pi}\xi^{i}}\right] + \tau\left[\frac{F(\xi^{r}) - \exp\{-j(\pi/4 + \xi^{r2})\}}{2\sqrt{\pi}\xi^{r}}\right]. \tag{30}$$

The diffracted field of a line source consists of two half-plane contributions. Therefore, the total field from the strip u_{sp}^{t} is given by

$$u_{sp}^{t}(\mathbf{r}) = u_{hp}^{t}(\rho_1, \phi_1) + u_{hp}^{t}(\rho_2, \phi_2) - u_{ext}(\mathbf{r}), \quad u_{ext} = U(\cos\phi/2)\{u^{i}(\xi^{i}) + u^{d}(\xi^{r})\} \tag{31}$$

where $F(x)$ is the Fresnel integral defined by

$$F(x) = \sqrt{j/x}\int_{x}^{\infty} \exp(-jt^2)dt \tag{32}$$

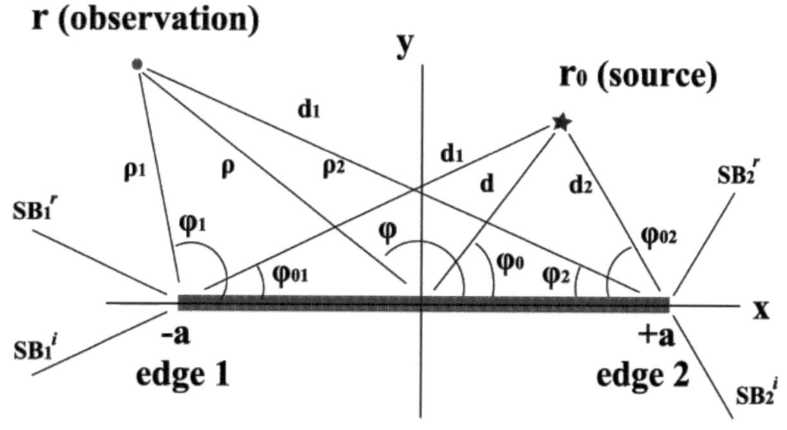

Figure 13.
Coordinates for scattering from a conducting strip.

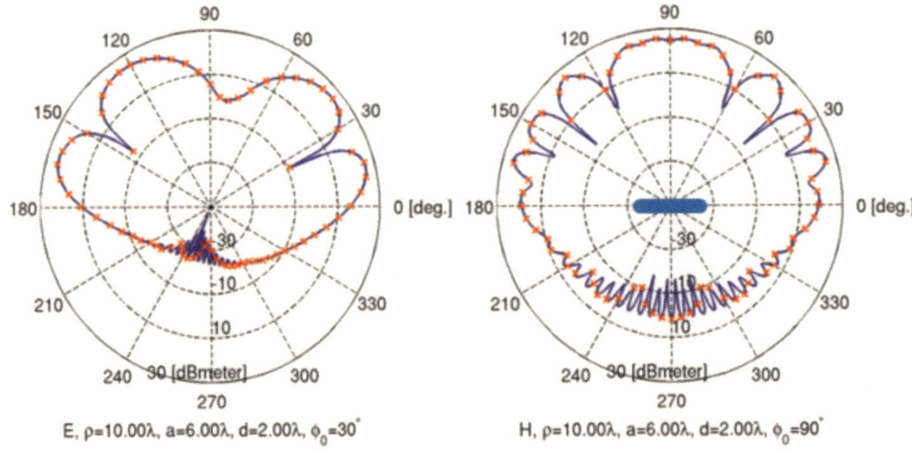

Figure 14.
Calculated near field of strip, left: E-pol., $\phi_o = 30^\circ$, right: H-pol., $\phi_o = 90^\circ$, symbol $\times$: exact solution.

and $U(x)$ is the Heaviside step function $U(x) = 1(x \geq 0); \ = 0(x < 0)$. In these equations, ξ^i and ξ^r are the detour parameters of the incident and reflected waves, respectively, given as

$$\xi^{i,r} = -\sqrt{\frac{4kd\rho}{d + \rho_{1,-1} + \rho}} \cos \frac{\phi^{i,r}}{2}. \tag{33}$$

In Eq. (29), u^d is the diffracted field associated with Keller's diffraction coefficients is expressed by the elementary function and τ is determined by the polarization, namely $\tau = -1$ for E-wave and $\tau = +1$ for H-wave. These formulas are available even if the observation point is on the shadow boundary (SB). For a two-face reflector, the existence of a double or single reflected wave depends on whether the reflection point is inside or outside the reflecting SB [10]. **Figure 14** is the near field RCS of a 6λ-long strip at observation $\rho = 10\lambda$ when a cylindrical wave is incident from $d = 2\lambda$, while **Figure 4** in the main body of this paper is the RCS of a $5\lambda \times 5\lambda$ corner reflector at $\rho = 15\lambda$ and $d = 10\lambda$. These results of the strip are checked against exact solutions (symbol $\times$ in **Figure 14**) and are in close agreement.

DOI: http://dx.doi.org/10.5772/intechopen.1011824

Author details

Hirokazu Kobayashi
Electromagnetic Wave System Laboratories, Tokyo, Japan

*Address all correspondence to: h.kobayashi@kobayashi-jim.com;
hirokazu.kobayashi.japan@gmail.com

References

[1] Kaatze U. Techniques for measuring the microwave dielectric properties of materials. Metrologia. 2010;**47**(2):91-113

[2] Clarke RN et al. A Guide to the Characterisation of Dielectric Materials at RF and Microwave Frequencies. London: Institute Meas. Control, National Phys. Lab; 2003

[3] Zivkovic I, Murk A. Free-Space Transmission Method for the Characterization of Dielectric and Magnetic Materials at Microwave Frequencies. London, UK: IntechOpen; 2012. DOI: 10.5772/51596

[4] Kawabata H, Hasuike K, Kobayashi Y, Ma Z. Multi-frequency measurements of complex permittivity of dielectric plates using higher order modes of a balanced-type circular disk resonator. In: 2006 36th Euro. Microwave Conf. pp. 388-391

[5] Mensa D. High Resolution Radar Cross-Section Imaging. London: Artech House; 1991

[6] Kobayashi H, Yamaguchi Y, Cui Y. Simple near-field to far-field transformation method using antenna Array factor. Journal of Wireless Networking and Communications. 2012; **2**:43-48

[7] Ahmad F, Zhang Y, Amin MG. Three-dimensional wideband beamforming for imaging through a Single Wall. IEEE Geoscience and Remote Sensing Letters. 2008;**5**(2):176-179

[8] Arfken G, Webwer H, Harris F. Mathematical Methods for Physicists. 7th ed. NY: Academic Press; 2013

[9] Kobayashi H, Inami M, Park S, Yamaguchi Y, Singh G, Cui Y. Radar imaging by using GTD near-field model and antenna Array-factor. In: 2012 International Symposium on Antennas and Propagation (ISAP2012). pp. 616-619

[10] Kobayashi H. Electromagnetic Wave in Space. Japan: Niigata University; 2011 (in Japanese)

[11] Pathak PH. Techniques for high-frequency problems. Ch. 4. In: Lo YT, Lee SW, editors. Antenna Handbook. NY: Van Nostrand Reinhold; 1988

[12] Daniels D. Surface-Penetrating Radar. In: 1966 IEE, Radar, Navigation and Avionics Series 6, UK

[13] Born M, Wolf E. Principles of Optics. 2nd ed. NY: Pergamon Press; 1964

[14] Kobayashi H, Chen KS, Yang Y. Theory of reflection/refraction by multilayer dielectric plates and its applications. In: 2025 IEICE Tech. Rep., EMT2025-29. 2025. pp. 103-108 (in Japanese)

[15] Kobayashi H. Physical optics. In: Yeap KH, Hirasawa K, editors. Analyzing the Physics of Radio Telescopes and Radio Astronomy. UK: IGI Global; 2020

Section 2

Antenna Design and Field Analysis

Chapter 3

Design and Analysis of Slotted Waveguide Antennas for 6G Applications

Jitendra Prajapati, Jyothishree Pillai and Madhur Deo Upadhayay

Abstract

The advent of 6G technology demands advanced antenna systems capable of meeting the requirements of ultra-high data rates, low latency, and massive connectivity. Slotted waveguide antennas, known for their high gain and precise beam-steering capabilities, present a promising solution. This chapter delves into the design, analysis, and optimization of slotted waveguide antennas tailored for 6G applications. It begins with a comprehensive review of existing literature, followed by the proposed design methodology. Numerical simulations and performance evaluations are presented to demonstrate the antenna's effectiveness in supporting 6G use cases. The findings underscore the potential of slotted waveguide antennas in addressing the challenges of next-generation wireless communication systems.

Keywords: 6G communication, antenna, beam scanning, electrical equivalent circuit, high gain, slotted waveguide

1. Introduction

With peak data rates of over 1 Tbps, submillisecond latency, and smooth physical and virtual environment integration, 6G technology is poised to transform wireless communications. These objectives demand the realization of novel antenna systems that can function in the sub-terahertz (THz) frequency range. Slotted waveguide antennas are unique among antenna technologies because of their high efficiency, scalability, and capacity to produce directed, narrow beams. The range of 6G networks will vary significantly depending on the frequency bands used and the deployment environment. It is anticipated that 6G will operate across a wide spectrum, including sub-THz and THz bands (100 GHz to 3 THz), mm-wave bands (30 GHz to 100 GHz), mid-bands (1 GHz to 30 GHz), and low bands ($<$ 1 GHz) [1]. Sub-THz and THz frequencies will primarily be used for high-speed data transfer and ultra-low-latency applications, but will have a short range, typically a few meters to hundreds of meters. With their higher propagation capability, micrometer wave bands can extend up to a few kilometers under line-of-sight (LOS) conditions and are suitable for high-density urban areas and backhaul networks. Mid-band frequencies are expected to balance coverage and capacity, with ranges extending to several kilometers, making them suitable for suburban and rural applications. Low-band frequencies offer wide coverage, reaching tens of kilometers, but with reduced data rates, making them ideal for rural connectivity and IoT applications. Advanced technologies like beamforming, MIMO, and dense network deployments will play a crucial role in optimizing the range and capacity of 6G networks. At the same time, propagation environments and transmission power will also significantly influence their performance. Antennas are the front-end components of communication. However, typical antennas are too small to achieve sub-Thz frequencies and suffer from propagation losses. Thus, waveguide antennas are often utilized at these frequencies. These are intended to steer electromagnetic waves through the waveguide construction with minimal losses. The perforations in the waveguide transmit electromagnetic waves between the waveguide and open space. Waveguide antennas, including all metal waveguides, dielectric loaded waveguides, and substrate-integrated waveguides, are designed to provide highly directed sub-THz frequency radiations [2]. All metal waveguides outperform printed antennas in terms of broadside radiation stability. The broadside radiation pattern of a waveguide antenna refers to a pattern in which the maximum radiation is perpendicular to the length of the waveguide. This pattern is advantageous in various applications for the following reasons:

- *High-directivity*: The broadside pattern focuses energy in a specific direction, making it ideal for applications requiring high gain and narrow beamwidth, such as radar and long-distance communication.
- *Enhanced coverage*: Broadside patterns provide uniform coverage over a defined area, which is useful for scanning or illuminating large regions without requiring complex steering mechanisms.
- *Efficiency in power utilization*: Energy is efficiently directed in the broadside direction, reducing power wastage in undesired directions, thereby improving overall system efficiency.
- *Compatibility with arrays*: Broadside waveguide antennas can be easily integrated into arrays to achieve higher gain, better beam steering, and adaptive radiation patterns.
- *Low side and back lobes*: With proper design, broadside antennas can minimize side and back lobes, reducing interference and improving performance in high-noise environments.
- *Suitable for planar applications*: The broadside pattern works well with planar structures, making it ideal for applications like phased arrays, 5G base stations, and modern communication systems.
- *Polarization control*: Broadside waveguide antennas offer flexibility in polarization (linear, circular, or elliptical), enhancing their adaptability to diverse applications.

1.1 Applications that require broadside radiation pattern

- *Wireless communications*: Provides high-gain links for point-to-point communication [1].
- *Satellite communications*: Ensures precise targeting of ground stations [2].

- *Unmanned aerial vehicles (UAVs) communication*: Broadside beam antennasin UAVs improve communication links with ground stations and satellites [3].

1.2 History and background of slotted waveguide antennas for 6G

Slotted waveguide antennas have been a cornerstone of antenna technology since their inception during World War II, primarily for radar systems. Invented to meet the demands of high-gain, narrow-beam radiation in military applications, they were initially used in aircraft and naval radar systems. The concept of introducing slots into a waveguide was revolutionary, as it allowed for controlled radiation of electromagnetic waves, making these antennas highly efficient and directive. Over the decades, slotted waveguide antennas have evolved significantly to meet emerging technological needs. Early designs focused on simple arrays of slots cut into rectangular waveguides, radiating at discrete frequencies. However, with advancements in materials and computational tools, more sophisticated designs emerged, allowing for complex beam-shaping, wide bandwidth, and polarization control. By the mid-twentieth century, these antennas were widely adopted in communication and radar systems, where high efficiency and durability were critical.

The development of 5G networks marked a significant shift in wireless communication, with a strong focus on higher frequencies, including millimeter waves. Slotted waveguide antennas proved instrumental in this transition due to their ability to operate efficiently at millimeter-wave frequencies while offering high gain and beam-steering capabilities. These attributes made them ideal for applications like fixed wireless access, massive MIMO systems, and satellite communications [4].

As the vision for 6G networks takes shape, the demand for even higher frequencies (terahertz range) and ultra-low-latency communication has brought slotted waveguide antennas back into the spotlight. The inherent advantages of waveguide structures, such as low loss and high power-handling capacity, make them suitable candidates for 6G systems. Additionally, their capability to support advanced features like dynamic beamforming, reconfigurability, and integration with metasurfaces aligns with 6G requirements for extreme data rates, ubiquitous connectivity, and energy efficiency. Advancements in materials and fabrication techniques are enabling waveguides and slots to operate effectively at terahertz frequencies, which are central to 6G communication.

Integration with Metasurfaces: Combining slotted waveguides with metasurfaces allows for reconfigurable beam patterns, improving coverage and adaptability in dynamic environments. Miniaturization and Integration: Compact designs suitable for integration with 6G devices, such as drones, satellites, and IoT nodes, are being explored. Enhanced designs focus on achieving high directivity while minimizing interference, critical for high-density 6G deployments.

Energy Efficiency: Optimizing power consumption and thermal management in slotted waveguide antennas is essential for sustainable 6G networks.

1.3 Methodological approach

In a slotted waveguide antenna array without phase shifters, beam scanning can be achieved through the following methods:

1.3.1 Mechanical movement

The entire antenna array or individual radiating elements can be physically rotated or tilted to steer the main beam direction. This movement can be achieved using mechanical actuators such as motors, gears, or pulleys. This method, however, makes the structure bulky and also consumes more power, and servo motor noise interference may also come into picture.

1.3.2 Slot orientation

By changing the orientation or shape of the slots in the waveguide structure, the radiation pattern of the antenna array can be altered, resulting in beam steering. This approach requires careful design and optimization of the slot configuration to achieve the desired beam scanning characteristics. Conversely, this method is less cumbersome, reduces self-interference, and causes very little performance degradation as a result of aging because it does not require extra circuitry or an external component, such as motors, phase shifters, etc., to provide phase shifting.

1.3.3 Frequency scanning

Another beam scanning method involves varying the antenna array's operating frequency. By tuning the frequency, the phase relationship between the radiating elements changes, resulting in beam steering. This approach, known as frequency scanning or frequency agility, can be implemented using tunable components such as varactor diodes or digitally controlled oscillators. This technique requires a larger bandwidth compared to other techniques.

1.3.4 Reflection grating

In some cases, a reflective grating can be used to steer the beam direction without physically moving the antenna elements. By adjusting the spacing or geometry of the grating elements, the phase of the reflected waves can be manipulated to achieve beam scanning.

1.4 Types of slotted waveguide antenna

1.4.1 Air-filled slotted waveguide antenna

An air-filled slotted waveguide antenna is a type of high-performance antenna that uses an air-filled waveguide as its transmission medium. The waveguide, typically constructed from a metal such as aluminum or brass, supports the propagation of electromagnetic waves while minimizing losses due to its low-loss air core. Slots are precisely cut into the waveguide's surface, often along the broad or narrow side, to serve as radiating elements. The placement, size, and orientation of these slots are carefully designed to control the antenna's radiation pattern, polarization, and impedance matching. Air-filled slotted waveguide antennas are widely used in applications requiring high gain, narrow beamwidth, and efficient power handling, such as radar systems, satellite communications, and military applications. Their design benefits from the simplicity of the air dielectric, which avoids the losses and dispersion associated with solid dielectric materials while also enabling operation at high frequencies, including the millimeter-wave range. These antennas offer excellent efficiency, durability, and thermal stability, making them well-suited for environments with stringent performance and reliability requirements [5–8].

1.4.2 Dielectric-filled slotted waveguide antenna

Dielectric waveguide antennas offer superior radiation efficiency compared to traditional waveguide-slot arrays [9, 10]. This improvement is attributed to the reduction of conductor resistance and radiation leakage losses associated with dielectric waveguides [11–13]. Furthermore, dielectric waveguides enable wider bandwidth operation, surpassing the capabilities of classical waveguide-slot arrays [14, 15]. The use of dielectric materials facilitates better impedance matching and supports broader operating frequency ranges. These materials also allow for the miniaturization and seamless integration of antenna elements, making them ideal for space-constrained applications [16, 17]. Additionally, dielectric waveguide antennas demonstrate lower cross-polarization levels than waveguide-slot arrays, effectively minimizing interference from undesired polarizations during signal transmission and reception [16, 18–21].

By optimizing the design of dielectric structures, desirable radiation characteristics, such as beam steering and shaping, can be achieved. Moreover, the simplified design and manufacturing processes of dielectric waveguide antennas contribute to lower production costs, especially significant for millimeter-wave antennas [22–24]. The transverse polarization in dielectric waveguide antennas pertains to polarization components perpendicular to the waveguide axis. This occurs when the transverse electromagnetic field components near the dielectric rod induce currents, generating waves polarized transversely to the waveguide axis.

This study focuses on comparing two types of dielectric waveguide antennas: one employing grooves in the dielectric and the other utilizing quarter-wavelength cylindrical pins, to evaluate their performance across key parameters. The grooved antenna produces longitudinal polarization, with the electric field aligned parallel to the waveguide axis [25, 26]. In contrast, the antenna with quarter-wavelength pins generates transverse polarization, with the electric field oriented perpendicular to the waveguide axis [27]. The grooved antenna exhibits a broadside radiation pattern, concentrating radiation perpendicular to the waveguide axis. Meanwhile, the quarter-wavelength pin antenna features an end-fire radiation pattern, directing radiation parallel to the waveguide axis.

Both antenna types achieve wide bandwidth; however, the grooved antenna enhances impedance matching and minimizes reflections, whereas the quarter-wavelength pin antenna delivers higher directivity, producing a focused beam with increased gain. The grooved structure simplifies fabrication, reducing manufacturing costs. Conversely, the quarter-wavelength pins enable precise control over beam steering and shaping, offering greater flexibility in radiation direction. The choice between these two antenna designs depends on application-specific requirements, including polarization, radiation pattern, bandwidth, and space constraints. Both designs bring unique advantages, making them suitable for diverse applications, such as wireless communication systems, radar, and satellite communications.

1.4.3 Substrate-integrated slotted waveguide antenna

A substrate-integrated slotted waveguide antenna (SISWA) is an innovative design that integrates waveguide structures into a planar dielectric substrate, enabling compact, lightweight, and cost-effective solutions for high-frequency applications.

These antennas leverage the concept of substrate-integrated waveguides (SIWs), which mimic traditional metallic waveguides but use metallized vias and parallel conductive plates within the substrate to confine electromagnetic waves. This design ensures efficient wave propagation while maintaining compatibility with printed circuit board (PCB) manufacturing techniques [28, 29].

In a SISWA, radiating slots are etched into the top or bottom conductive layer of the SIW. These slots are carefully designed and positioned to achieve specific radiation characteristics, such as desired polarization, beam direction, and impedance matching. The slots typically resonate at particular frequencies, allowing the antenna to operate efficiently within predefined bands. By adjusting the slot dimensions and locations, designers can control the radiation pattern and enhance the bandwidth or gain of the antenna.

One of the key advantages of SISWAs is their planar structure, which allows for easy integration with other components in microwave and millimeter-wave circuits, such as filters, amplifiers, and power dividers. Additionally, the use of a substrate reduces fabrication costs and makes the antenna compatible with mass production methods. The dielectric material in the substrate also provides mechanical support and protects the antenna from environmental influences.

SISWAs are widely used in applications such as 5G communication, satellite systems, automotive radar, and IoT devices. They offer high performance, including low insertion loss, high gain, and efficient radiation, while maintaining a small form factor. This combination of features makes substrate-integrated slotted waveguide antennas a popular choice for modern wireless communication systems, where compactness, efficiency, and integration are crucial. The foundational work reported in 2015 [30] introduced a probe-fed Substrate-Integrated Waveguide Cavity-Backed Slot (SICBS) antenna to develop a four-element Multiple-Input Multiple-Output (MIMO) system operating in the WLAN band (2.4–2.485 GHz). The antenna unit, designed to radiate in the broadside direction, incorporates a rectangular slot etched into a square-shaped SIW cavity. Four such antenna units were closely arranged by rotating them orthogonally to each other [31]. A parametric study highlighted the impact of orientation and spacing between the antenna elements on isolation performance. A compact, closely spaced MIMO antenna that uses substrate-integrated waveguide (SIW) technology, offering improved isolation and integration capabilities, is demonstrated in [32]. In [33] waveguide structures are introduced with magnetic walls derived from mushroom-shaped metamaterials, showcasing enhanced field confinement and novel electromagnetic behaviour. The analysis of radiation loss associated with edge-coupled coplanar waveguide (ECCPW) gap discontinuities, which contribute to the accurate characterization and optimization of such planar transmission lines, is investigated in [34].

2. Mathematical approach on slotted waveguide antenna

The mathematics behind a slotted waveguide antenna involves the analysis and design of wave propagation, radiation, and coupling within the waveguide [35]. Below are the key mathematical aspects:

2.1 Wave propagation in the waveguide

To get the wave behavior inside a rectangular waveguide, Maxwell's equations are solved using the following boundary conditions:

- The tangential component of the electric field must be zero on the conducting walls.
- The normal component of the magnetic field must be zero on the conducting walls.

For a rectangular metallic waveguide, under source-free condition, the wave equation results in Helmholtz's equation and can be written as:

$$\nabla^2 E + k^2 E = 0 \tag{1}$$

where E is the electric field, and $k = \frac{2\pi}{\lambda}$ is the wavenumber for which λ determines the wavelength. Similarly, for the magnetic field H, the wave equation takes the form:

$$\nabla^2 H + k^2 H = 0 \tag{2}$$

where H is the magnetic field, and $k = \frac{2\pi}{\lambda}$ is the wavenumber for which λ determines the wavelength. *Modes*: The waveguide supports specific modes, typically TE_{mn} or TM_{mn}, characterized by their cutoff frequency:

$$f_c = \frac{c}{2}\sqrt{\left(\frac{m}{a}\right)^2 + \left(\frac{n}{b}\right)^2} \tag{3}$$

where c is the speed of light, a and b are the dimensions of the waveguide, and m, n are mode indices.

2.2 Slot design and coupling

Slot spacing: To achieve constructive interference in the broadside direction, the spacing between slots (d) is:

$$d = \frac{\lambda_g}{2} \tag{4}$$

where λ_g is the guided wavelength:

$$\lambda_g = \frac{\lambda}{\sqrt{1 - \left(\frac{f_c}{f}\right)^2}} \tag{5}$$

Slot length and width: The length (l) of the slots determines the coupling strength and is typically around half the guided wavelength:

$$l \approx \frac{\lambda_g}{2} \tag{6}$$

Slot displacement: For inclined slots, the displacement from the centerline introduces a phase shift and affects the radiation pattern:

$$\Delta\phi = \beta \cdot x \tag{7}$$

where $\beta = \frac{2\pi}{\lambda_g}$ is the phase constant, and x is the displacement.

2.3 Radiation pattern

Array factor: The radiation pattern of the antenna is a product of the array factor and the element factor. For N slots, the array factor is:

$$AF(\theta) = \sum_{n=1}^{N} a_n e^{j\beta_n d_n \cos\theta} \tag{8}$$

where a_n is the amplitude, β_n is the phase shift, and d_n is the slot spacing.

Broadside radiation: For maximum radiation in the broadside direction ($\theta = 90^\circ$):

$$\beta_n d_n = 2m\pi, \quad m \in \mathbb{Z}. \tag{9}$$

Main beam direction: The main beam direction occurs at:

$$\theta_m = \cos^{-1}\left(\frac{m\lambda}{Nd}\right) \tag{10}$$

where m is an integer representing the order of the grating lobe (for the main lobe, $m = 0$).

Half-Power Beamwidth (HPBW): The Half-Power Beamwidth (HPBW) can be approximated for a uniform array as:

$$\text{HPBW} \approx \frac{50.6\lambda}{Nd} \tag{11}$$

where N is the number of slots, d is the slot spacing, and λ is the wavelength at the operating frequency.

First Null Beamwidth (FNBW): The First Null Beamwidth (FNBW) is given by:

$$\text{FNBW} = \frac{2\lambda}{Nd} \tag{12}$$

2.4 Impedance matching

The impedance of each slot (Z_s) affects the overall input impedance of the waveguide. The input impedance is calculated as:

$$Z_{\text{in}} = Z_0\left(1 + \frac{Z_s}{Z_0}\right) \tag{13}$$

where Z_0 is the characteristic impedance of the waveguide. Matching is achieved by optimizing slot size, position, and orientation.

2.5 Power distribution in slotted waveguide antenna Array

The power distribution in a slotted waveguide antenna array depends on the slot arrangement, the excitation coefficients, and the aperture field distribution. The following equations provide a mathematical description:

Total radiated power: The total power radiated by the array is given by:

$$P_{\text{total}} = \int_0^{2\pi} \int_0^{\pi} U(\theta, \phi) \sin\theta d\theta d\phi \tag{14}$$

where:

- $U(\theta, \phi)$: Radiation intensity as a function of elevation (θ) and azimuth (ϕ) angles.
- $\sin\theta$: Element factor for spherical coordinate integration.

Power radiated by individual slot: The power radiated by the n-th slot in the array is:

$$P_n = \frac{1}{2} |I_n|^2 R_n \tag{15}$$

where:

- I_n: Current amplitude at the n-th slot.
- R_n: Radiation resistance of the slot.

Power distribution along the array: For a waveguide carrying forward-traveling power, the power radiated at the n-th slot is proportional to the local field amplitude:

$$P_n = P_{\text{input}} \left(1 - e^{-2\alpha(n-1)d}\right) \tag{16}$$

where:

- P_{input}: Input power at the waveguide port.
- α: Attenuation constant of the waveguide.
- d: Distance between consecutive slots.

Excitation coefficients: The excitation coefficients are used to control the power distribution and are defined as:

$$a_n = \frac{I_n}{I_1} = e^{-\alpha(n-1)d} \tag{17}$$

where I_1 is the current at the first slot.

Normalized power distribution: The normalized power radiated by the n-th slot, relative to the total power, is:

$$P_n^{\text{norm}} = \frac{P_n}{P_{\text{total}}} = \frac{|I_n|^2 R_n}{\sum_{i=1}^{N} |I_i|^2 R_i} \tag{18}$$

2.6 Efficiency and gain

Antenna efficiency:

$$\eta = \frac{P_r}{P_{\text{in}}} \tag{19}$$

Gain:

$$G = \eta \cdot \frac{4\pi A_e}{\lambda^2} \tag{20}$$

where A_e is the effective aperture area of the antenna.

These equations form the foundation for the design and optimization of slotted waveguide antennas. Advanced numerical methods, such as finite element analysis (FEA) or method of moments (MoM), are often employed for precise simulations and validations.

3. Equivalent circuit model of a generalized slotted waveguide antenna

The equivalent circuit of a generalized slotted waveguide antenna can be represented as a combination of the waveguide impedance, slot impedance, and transformer for coupling to the network. The equivalent circuit of a generalized slotted waveguide antenna can be represented using the following components:

3.1 Input impedance of the waveguide

The port impedance of the waveguide is given by:

$$Z_g = \frac{Z_0}{\sqrt{1 - \left(\frac{f_c}{f}\right)^2}} \tag{21}$$

where: $Z_0 = \sqrt{\frac{\mu}{\epsilon}}$ is the characteristic impedance of the waveguide, and f_c is the cutoff frequency.

3.2 Slot impedance

Each slot on the waveguide can be modeled as a parallel LC circuit, where the admittance of the slot is:

$$Y_P = j\omega C + \frac{1}{j\omega L} \tag{22}$$

Here:

- C: Capacitance of the slot.
- L: Inductance of the slot.

3.3 Scattering parameter

The reflection coefficient (S_{11}) of the waveguide is given by:

$$S_{11} = \frac{1}{2(1 + Y_P Z_g)} \tag{23}$$

3.4 Transformer model

The transformer ratio, relating the waveguide impedance (Z_g) [29] to the slot impedance (Z_s), is defined as:

$$n = \frac{Z_g}{Z_s} \tag{24}$$

where n is the transformer ratio.

3.5 Additional components for bandwidth balancing

To balance the bandwidth, an equivalent parallel RL combination can be added before the slot, with:

- R_1: Resistance.
- L_1: Inductance.

3.6 Termination for free space radiation

The step-down transformer is terminated with a high resistance (R_4), approximating an infinite resistor to model free space.

3.7 Impedance of the slot

The impedance of the slot is calculated as:

- *R_s* = Calculated based on radiation properties and slot dimensions.
- *Z_g*: Waveguide impedance, dependent on the operating frequency.

$$Z_g = \frac{Z_o}{\sqrt{1 - \left(\frac{f_c}{f}\right)^2}} \tag{25}$$

- T: Transformer to couple the waveguide slots to the external network.
- L_s and C_s: Parallel LC combination representing the slot's equivalent model.
- R_3: Slot resistance, determined by the waveguide and slot dimensions.
- R_∞: Equivalent termination for radiation into free space.

For instance, the slotted waveguide antenna array composed of hexagonal slot embedded in a linearly placed vertical slot arrays represented in the work of J. Pillai et al. [8] can be represented as given in **Figure 1a** and **b**. The slots on the waveguide are modeled as a parallel LC combination, and they are coupled to the network through a transformer. The transformer ratio, defined as the ratio $\frac{Z_g}{Z_s}$, is calculated to be 0.775. **Figure 2** illustrates the equivalent circuit model's frequency response, which

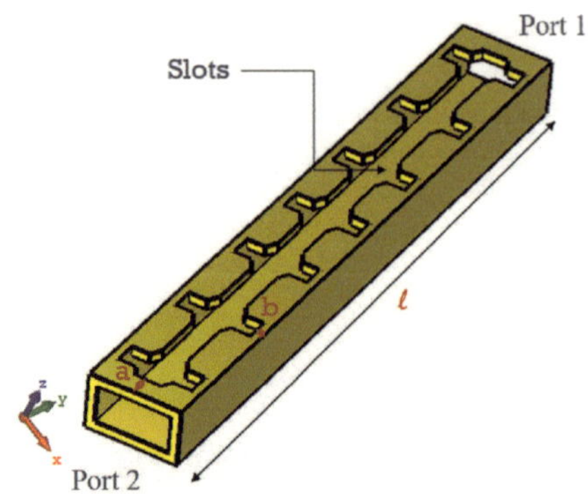

(a)

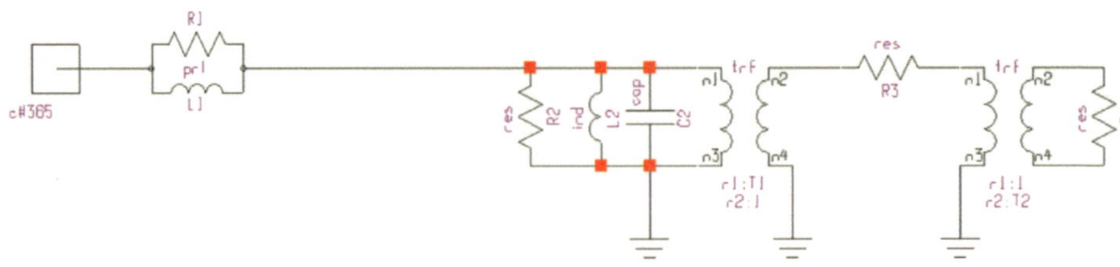

(b)

Figure 1.
(a) The slotted waveguide antenna proposed in the work of J. Pillai et al. [8], and (b) Equivalent LCT model of the slotted waveguide.

is validated using SPICE software. The resonance observed in **Figure 2** is consistent with the frequency response of the proposed structure.

The calculated inductance and capacitance values are $L_2 = 12.5$ nH and $C_2 = 3.2$ pF, respectively. Since the objective is to ensure that the waveguide radiates into free space, the step-down transformer is terminated with an infinite resistance value R_4.

The slot on the waveguide is modeled by computing the impedance value, given as $R_3 = 272.51\Omega$. To optimize the bandwidth, an equivalent parallel RL combination is incorporated with component values $R_1 = 30\Omega$ and $L_1 = 0.04$ nH.

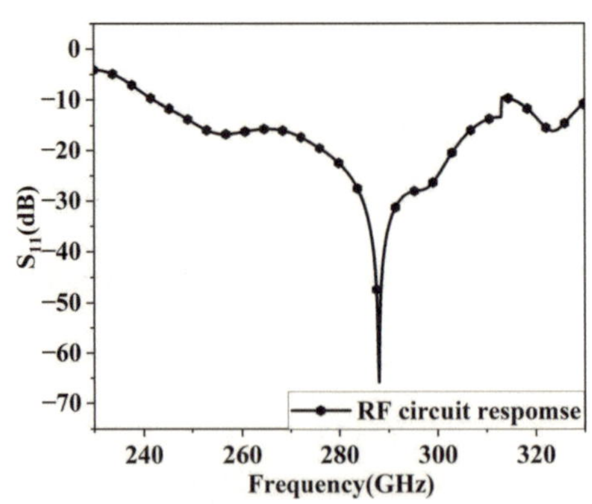

Figure 2.
Frequency response of the equivalent model.

4. Array factor and electric field distribution of a generalized slotted waveguide antenna system

The array factor of a slotted waveguide antenna array represents the combined radiation pattern resulting from the interference of fields from multiple slots, influenced by their spacing, amplitude, and phase. The electric field distribution describes the spatial variation of the radiated field, determined by the individual slot characteristics and the array geometry.

4.1 Array factor (AF)

$$AF(\theta, \phi) = \sum_{n=1}^{N} I_n e^{j(kdn \sin(\theta) \cos(\phi) + \beta n + \psi_n)} \tag{26}$$

Where:

- I_n = Amplitude of the nth slot
- $k = \frac{2\pi}{\lambda}$ = Wavenumber
- d = Spacing between slots
- β = Phase shift due to waveguide propagation
- ψ_n = Additional phase of the nth slot
- N = Number of slots
- θ = Elevation angle
- ϕ = Azimuth angle

4.2 Electric field distribution ($E(\theta, \phi)$)

$$E(\theta, \phi) = E_0 F(\theta, \phi) \cdot AF(\theta, \phi) \tag{27}$$

Where:

- E_0 = Peak electric field amplitude
- $F(\theta, \phi)$ = Element factor (determined by the slot's shape and orientation)
- $AF(\theta, \phi)$ = Array factor (as defined above)

Again, for instance, if we take a series-fed slotted antenna array where each element is arranged at a distance "d", i.e., $\frac{\lambda}{4}$ and is connected to each other by means of a high impedance slot as in the work of J. Pillai et al. [8]. Then net electric field can be written as: -

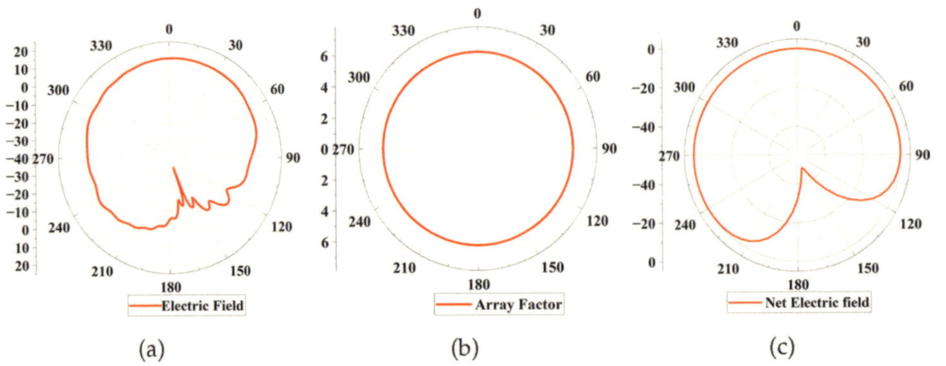

Figure 3.
Analytically calculated electric field distribution of (a) Single element, (b) Array factor, and (c) Net Electric field due to all elements.

$$E = E_1 + E_1 e^{j\theta} e^{j2\pi d/\lambda} + E_1 e^{j2\theta} e^{j4\pi d/\lambda} + \dots\dots\dots\dots E_1 e^{j6\theta} e^{j12\pi d/\lambda} \tag{28}$$

Array Factor for the proposed structure

$$AF = e^{j\left[\frac{N-1}{2}\right]\left(\frac{\theta}{2}+\frac{\pi}{4}\right)\frac{\sin\left[\left(\frac{\theta}{2}+\frac{\pi}{4}\right)N\right]}{\sin\left[\frac{\theta}{2}+\frac{\pi}{4}\right]}} \tag{29}$$

The theoretically calculated slotted waveguide single element electric field, array factor, and net electric field is shown in **Figure 3**.
By curve fitting analysis, the E-Field of a single element is given by the Gaussian function

$$y = y_0 + Ae^{-\frac{(x - x_c)^2}{2\omega^2}} \tag{30}$$

where y_0 =14.088 (offset value of y when x is far away from the central point x_c), Mean $x_c = 160$; where, Mean is deduced to be $\frac{1}{(n-1)} \frac{L\pi}{\lambda}$, for n = 7 elements.

Standard deviation $\omega = 22.2$ and A = -47.98 (Peak value of the Gaussian curve) (**Figure 3**).

5. Conclusions

The evolution of slotted waveguide antennas aligns seamlessly with the stringent requirements of 6G wireless communication systems. Their high-gain, precise beam-steering capabilities and ability to support advanced features like dynamic beamforming and reconfigurability make them a pivotal technology in addressing the challenges of ultra-high data rates, low latency, and massive connectivity. The integration of slotted waveguide antennas with metasurfaces and advancements in material science has extended their operability to sub-THz and THz frequencies, enhancing their adaptability and efficiency for next-generation networks. The analysis of frequency bands highlights the versatility of 6G systems in catering to diverse deployment scenarios—from wide-area coverage in rural regions using low-band frequencies to high-speed data transfer in urban environments leveraging sub-THz and THz bands. Slotted waveguide antennas complement these advancements by providing robust performance across these spectral ranges, particularly under line-of-sight conditions. Through the design, simulation, and optimization efforts presented, this chapter demonstrates that slotted waveguide antennas not only meet but exceed the requirements for 6G applications. Their proven efficiency, coupled with new innovations in fabrication and integration, underscores their potential as a cornerstone technology for the future of wireless communications.

Conflict of interest

This declaration is "not applicable".

Author details

Jitendra Prajapati[1*], Jyothishree Pillai[2] and Madhur Deo Upadhayay[1]

1 Shiv Nadar Institution of Eminence Deemed to be University, Delhi, Uttar Pradesh, India

2 JSS Academy of Technical Education, Noida, Uttar Pradesh, India

*Address all correspondence to: jitendra.prajapati@snu.edu.in

References

[1] Aqlan B et al. Sub-THz circularly polarized horn antenna using wire electrical discharge machining for 6G wireless communications. IEEE Access. 2020;**8**:117245-117252

[2] Martinez-de-Rioja E, Martinez-de-Rioja D, López-Sáez R, Linares I, Encinar JA. High-efficiency polarizer reflectarray antennas for data transmission links from a CubeSat. Electronics. 1802;**2021**:10

[3] Dicandia FA, Fonseca NJG, Bacco M, Mugnaini S, Genovesi S. Space-air-ground integrated 6G wireless communication networks: A review of antenna technologies and application scenarios. Sensors. 2022;**22**:3136. DOI: 10.3390/s22093136

[4] Hirokawa J, Zhang M. Waveguide slot array antennas. In: Chen Z, Liu D, Nakano H, Qing X, Zwick T, editors. Handbook of Antenna Technologies. Singapore: Springer; 2016. DOI: 10.1007/978-981-4560-44-351

[5] Siegel PH. Terahertz technology. IEEE Transactions on Microwave Theory and Techniques. 2002;**50**(3):910-928. DOI: 10.1109/22.989974

[6] Lioubtchenko DV, Tretyakov SA, Dudorov SN. Millimeter-wave waveguides. New York, NY: Springer; 2003. DOI: 10.1007/b105858

[7] El Misilmani H, Husseini M, Kabalan K. Design of slotted waveguide antennas with low sidelobes for high power microwave applications. Progress In Electromagnetics Research C. 2015;**56**: 15-28. DOI: 10.2528/PIERC14121903

[8] Pillai J, Jitendra P, Deo UM. Efficiency enhancement of a circularly polarised slotted waveguide antenna array for sub-THz applications. Optical and Quantum Electronics. 2024;**56**. Artticle no. 805. DOI: 10.1007/s11082-024-06454z

[9] Shaaban MN, Nasybullin AR, Sedelnikov YE. Design and analysis antennas of transverse polarization on the dielectric waveguide. Journal on Wireless Communications and Networking. 2024;**2024**:14. DOI: 10.1186/s13638-024-02342-y

[10] Torabi Y, Dadashzadeh G, Lalbakhsh A, Oraizi H. High-gain and low-profile dielectric-image-line leaky-waveantenna for wide-angle beam scanning at sub-thz frequencies. Optics and Laser Technology. 2022;**150**:107968

[11] Sahoo MC, Patani A. Slotted rectangular dielectric resonator antenna for the application of satellite communication. Wireless Personal Communications. 2023;**130**(2):837-855

[12] Adhikari R, Sbeah Z, Gupta R, Chauhan D, Nunzi J-M, Dwivedi RP. Compact and sensitive h-shaped metal-dielectric-metal waveguide plasmonic sensor. Plasmonics. 2022;**17**(4):1593-1606

[13] Li Y, Liao S, Xue Q, Che W. Transmission characteristics of flexible low-loss solid circular polymer dielectric waveguides for sub-thz applications. Journal of Infrared, Millimeter, and Terahertz Waves. 2023;**44**(1–2):110-133

[14] Sakakibara K et al. A linearly-polarized slotted waveguide array using reflection-cancelling slot pairs. IEICE Transactions on Communications. 1994; **77**(4):511-518

[15] Seki H, Goto N. Synthesis of circular polarization with nonresonant slots in

the narrow wall of a rectangular waveguide. Electronics and Communications in Japan (Part I: Communications). 1981;**64**(9):70-78

[16] Ji Z, Wang KX, Wong H. Circularly polarized dielectric rod waveguide antenna for Millimeter-wave applications. IEEE Transactions on Antennas and Propagation. 2018;**66**(10): 5080-5087. DOI: 10.1109/TAP.2018.2858182

[17] Generalov AA et al. Wide band mm- and sub-mm-wave dielectric rod waveguide antenna. IEEE Transactions on Terahertz Science and Technology. 2014;**4**(5):568-574

[18] Gao J, Li T, Wang H, Lei X, Wang K. A compact dual-band dual-linearly polarized waveguide slot array antenna with groove at waveguide bottom. International Journal of Microwave and Wireless Technologies. 2023;**15**(7): 1205-1211

[19] You Q, Wang Y, Huang M, Huang J, Zheng Z-W, Yunlong L. Wideband dual-polarized hollow-waveguide slot array antenna. IEEE Transactions on Antennas and Propagation. 2022;**70**(10):9326-9336

[20] Josefsson L, van't Klooster CGM. Dual polarised slotted waveguide SAR antenna. IEEE Antennas and Propagation Society International Symposium. 1992;**1**:625-628

[21] Qu Y, Ma J,Wang J, Zheng Q. Wideband ridged waveguide slot array antenna with low cross-polarization for satellite communications. In: 2024 IEEE International Conference on Computational Electromagnetics (ICCEM). IEEE. 2024. pp. 1-3. DOI: 10.1109/ICCEM60619.2024.10559117

[22] Van Trinh T, Park J, Song CM, Song S, Hwang KC. A 3-D metal-printed dual-polarized ridged waveguide slot array antenna for X-band applications. Applied Sciences. 2023;**13**:4996. DOI: 10.3390/app13084996

[23] Alsirhani K, Abdalmalak KA, Lee CS, Santamaría-Botello G, Segovia-Vargas D, García-Muñoz LE. Dielectric resonator antenna fed by tapered dielectric rod waveguide for 5G mm-wave applications. In: IEEE International Symposium on Antennas and Propagation and North American Radio Science Meeting. IEEE; 2020. pp. 149-150. DOI: 10.1109/IEEECONF35879.2020.9329488

[24] Pousi JP, Lioubtchenko DV, Dudorov SN, Raisanen AV. High permittivity dielectric rod waveguide as an antenna array element for millimeter waves. IEEE Transactions on Antennas and Propagation. 2010;**58**(3):714-719. DOI: 10.1109/TAP.2009.2039314

[25] Wu Y, Hirokawa J, Tomura T. Dual-polarized full-metal perpendicular-corporate-feed parallel-plate waveguide slot array antenna. IEEE Transactions on Antennas and Propagation. 2024;**72**(7): 6093-6098. DOI: 10.1109/TAP.2024.3356788

[26] Zhou S-G, Huang G-L, Chio T-H, Yang J-J, Wei G. Design of a wideband dual-polarization full-corporate waveguide feed antenna array. IEEE Transactions on Antennas and Propagation. 2015;**63**(11):4775-4782

[27] Deslandes D, Wu K. Design consideration and performance analysis of substrate integrated waveguide components. In: Proceedings of the 32nd European Microwave Conference (EuMC), Milan, Italy, 27–29 September 2002. Piscataway, NJ, USA: IEEE; 2002. pp. 23-26

[28] Chen Y et al. Broadband high-gain SIW horn antenna loaded with tapered

multistrip transition and dielectric slab for mm-wave application. IEEE Transactions on Antennas and Propagation. 2022;**70**(7):5947-5952. DOI: 10.1109/TAP.2022.3161349

[29] Wang L, Liao Q. Wideband multibeam SIW horn array with high beam isolation and full azimuth coverage. IEEE Transactions on Antennas and Propagation. 2021;**69**(9): 6070-6075. DOI: 10.1109/ TAP.2021.3069564

[30] Kumar A. Substrate integrated waveguide cavity-backed slot antenna with low cross-polarization over the full bandwidth. Microwave and Optical Technology Letters. 2024;**66**(1):e34019

[31] Sung Y. Closely spaced MIMO antenna based on substrate-integrated waveguide technology. Microwave and Optical Technology Letters. 2018;**60**: 1794-1798. DOI: 10.1002/mop.31249

[32] Yelizarov AA, Nazarov IV, Skuridin AA, Zakirova EA. Waveguide Structures with Magnetic Walls on the Basis of the Mushroom-shaped Metamaterials. In: 2020 IEEE International Conference on Actual Problems of Electron Devices Engineering (APEDE). IEEE; 2020. pp. 175-179

[33] Kumar M, Gowr R. Characterization and evaluation of radiation loss factor of edge coupled coplanar waveguide gap discontinuity. In 2016 Second International Conference on Computational Intelligence and Communication Technology (CICT). IEEE; 2016. pp. 415-418. DOI: 10.1109/ CICT.2016.88

[34] Kishk AA, Yakovlev AB, Glisson AW. Load-independent equivalent circuit model for transverse waveguide slots. In: 2005 IEEE Antennas and Propagation Society International Symposium. Vol. 4B. IEEE; 2005. pp. 310-313. DOI: 10.1109/ APS.2005.15528 09

[35] Josefsson L, Rengarajan SR. Slotted Waveguide Array Antennas: Theory, Analysis and Design. The Institution of Engineering and Technology. 2018. DOI: 10.1049/SBEW517E

Chapter 4

Image-Based Near-Field to Far-Field Transformation

Hirokazu Kobayashi

Abstract

This chapter discusses the theory of near-field to far-field transformation (NFFFT) for the radar cross section (RCS) of a radar target such as an aircraft. The RCS is one of the most important metrics used in the design and evaluation of radar and electronic warfare systems. When the electrical size of the target under test is large, the probe antenna measuring the waves scattered by the object has to be placed far enough away from the target to illuminate the plane wave. If the frequency and wavelength are f and λ, respectively, and the maximum dimension of the target is D, the far distance R is generally evaluated as $R > 2D^2/\lambda$. For example, when $D = 3$ m and $f = 10$ GHz, $R = 600$ m is an unrealistic condition. It is therefore conceivable that some method might be needed to theoretically transform the near-field scattering of the target into a far-field. As in the approach of this chapter, when a radar image is obtained, it is assumed that a far-field can be evaluated by considering the image as a secondary wave source. Based on this concept, the relationship between the radar image and the three-dimensional (3D) RCS can be derived. The theoretical validity is proved by numerical calculations by comparison with actual measurement data. The interesting result is that the unwanted waves can be suppressed by proper processing of the images containing the unwanted waves. The equivalence between time-domain processing, such as fast Fourier transform (FFT), and spatial-domain processing in image data is confirmed.

Keywords: electromagnetic near-field, far-field transformation, radar cross section (RCS), inverse synthetic aperture radar (ISAR), focusing, cylindrical scanning, antenna pattern, equivalence of time- and space-domain

1. Introduction

It is often difficult to directly measure RCS of an object (radar target) with a large electrical length in the conventional far-field measurement, because RCS is defined at such a distance that the illuminated wave is considered to be a plane wave [1]. Two main solutions have been adopted to address this situation. One is the compact range method, in which the object is placed in a phase-coherent electromagnetic field in the aperture area of a large parabolic reflector, and it has already been used for RCS and antenna measurements since the 1980s. The other is the near-field to far-field transformation (NFFFT) method, which acquires the electromagnetic near-field and mathematically analyzes the far-field from the near data. This was researched and developed as an antenna measurement theory at the same time as the compact range method; however, it was difficult to apply to RCS evaluation, which is a two-way measurement. To solve this problem, a method was proposed around 2000 to calculate the far-field RCS from radar images obtained by inverse synthetic aperture radar (ISAR) processing [2].

The above radar image can be considered as an equivalent wave source distribution of the object and it can be assumed that the scattering information of the target is included [3]. If a radar image can theoretically be constructed as a set of small wave sources, the far-field RCS of the scatterer can be evaluated [4–6]. In normal applications of ISAR/SAR, the distance between the radar and the target object is composed of far-field regions and there is no practical problem. However, NFFFT using radar images discussed in this chapter is a method to obtain measurement data in a near-field region [7, 8]. As a result, two major problems are expected. One is the problem that the distance between the near-field probe antenna and the target changes significantly, resulting in a large intensity difference in the resulting measurement data. This appears as a (focus) blur in the radar image, and it is thought that a large error will occur in the far-field RCS generated from the radar image. In order to overcome this problem, the authors assumed and rigorously formulated a small conductive sphere as the target [9]. In the far-field transformation, this acts as an automatic weighting according to the variable coordinates of the integral process and can be seen as an equivalent focal length correction.

Another issue is how the results are affected by multiple reflections, which are a unique phenomenon in the near-field but not in the far-field. For example, it is easy to understand this by using the ray theory of geometric optics to consider a two-faced dihedral corner reflector, which consists of two flat plates crossing at right angles. If the radar's transmitting (Tx) and receiving (Rx) antennas are far away from this target, the direction of the wave (ray) emitted by the Tx antenna and the wave reflected twice by the two orthogonal flat plates and returning to the radar are parallel to each other. This situation is difficult to be reproduced in the NFFFT measurements. In the near-field, the incident and reflected rays are not parallel, which explains why the NFFFT of objects that generate multiple reflections may not be able to accurately measure RCS. Taking an aircraft as an example, these multiple reflections correspond to resonant reflections caused by an air intake or when multiple flat parts are perpendicular to each other [10, 11].

The NFFFT has been studied for antenna measurements with large electrical lengths. The integral equation method is a traditional method that formulates strict relational expressions according to the probe scanning system [12] and is the basis of current commercial software using antenna NFFFT. RCS measurement is not possible with NFFFT using integral equations specialized for antenna measurement. Therefore, NFFFT based on Physical Optics (PO) [13] and Array Factor [14] has been proposed, but both require plane wave illumination, which is an obstacle to practical use. The transformation method discussed in this chapter is a method in which the illuminating wave is a non-plane wave, and the far-field is calculated from the ISAR image obtained in this case. Here, we formulate it for cylindrical scanning including the circumference; however, it is theoretically possible to formulate it for spherical scanning.

By replacing either the Tx or Rx port of RCS measurement with the input port of the antenna, the antenna near-field and image data can be obtained. If NFFFT processing is carried out simply, it can be assumed that the far-field is calculated from the antenna near-field image. This corresponds to the so-called far radiation pattern of the antenna. When confirmed with a pyramidal horn antenna, the transformed pattern was found to be almost identical to the far-field calculation value of the horn [15]. Measuring RCS or antenna of a large object requires considerable equipment, so it is very cost effective to be able to measure RCS and antennas with NFFFT described here in a relatively small anechoic chamber. Due to space limitations, this antenna measurement is not discussed further in this chapter.

In order to smoothly perform measurement and transformation processing in a unified manner, it is also necessary to develop a software code that performs a series of measurement and calculation processing. This code also incorporates a theoretical model that gives the near-field of spheres, plates, etc., and is capable of pre-evaluating the adjustment accuracy of NFFFT, etc., and post-processing using FFT for the purpose of removing unnecessary waves on the time-domain. Furthermore, it is shown that for multiple reflections within the target, the time-domain spurious wave suppression by FFT is found to be equivalent to the image space spurious wave removal processing by measuring the two-faced corner reflector with/without metal plate.

2. Theory of NFFFT in circular scanning

Figure 1 shows the coordinate system of 2D circular scanning. Microwave signals are transmitted and received by a probe antenna rotating 360 degrees in azimuth with radius ρ_0. In cylindrical scanning, it also moves in the z-axis direction. The inner area indicated by Ω in the same figure is the image generation area, which is smaller than the probe scanning area of radius ρ_0 and encloses the object to be measured with the maximum dimension D ($D \leq 2\rho_{\max} \leq 2\rho_0$). The directivity of the probe antenna shown in the figure is given by $P(\beta)$, where $P(\beta)$ is the angle measured from the antenna boresight axis, and its boresight direction is toward the center of the rotation axis. For example, when $P(\beta) = 1$ or $\cos\beta$ denotes the electric dipole perpendicular or tangential to the scanning plane, respectively. In addition, the Gaussian beam antenna whose half-width is given by β_0 is represented by $P(\beta) = \exp\left\{-\ln 2(\beta/\beta_0)^2\right\}$.

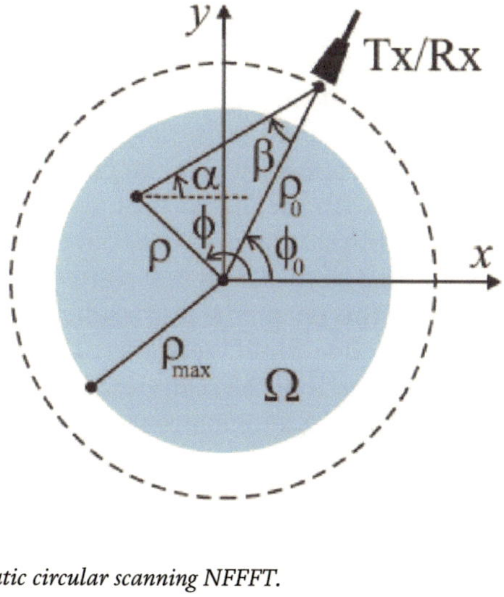

Figure 1.
Coordinate system in mono-static circular scanning NFFFT.

First, the measurement calibration is explained. A scattering object for calibration is placed in the image generation area at the center of the rotation axis. Many objects with known theoretical values, such as a conductive sphere, a plate, and a corner reflector, are used for the shape. Now, assume a relatively small size scatterer. Let $E_0^s(k)$ be the received signal ($k = 2\pi f/c, f$: measurement frequency, c: speed of light), we consider a general expression in the far-field:

$$E_0^s(k) = Q(k)A_0(k)\frac{\exp(-2jk\rho_0)}{2\sqrt{\pi}\rho_0^2}. \tag{1}$$

In the above equation, the frequency dependent factor $Q(k)$ is a quantity proportional to the field level of the Tx antenna, and $A_0(k)$ is a quantity related to the backscatter from the scatterer. The $A_0(k)$ is a quantity proportional to RCS and $\sigma = |A_0(k)|^2$. For example, an electrically small conducting sphere of radius a has $A_0(k) = 3\sqrt{\pi}k^2a^3$, an electrically large arbitrary shape with a projected cross-sectional area of S in the vertical incidence direction [1], $A_0(k) = -jkS/\sqrt{\pi}$. In the calibration process, $Q(k)$ of the Tx antenna takes the inverse of Eq. (1) and is given by:

$$Q(k) = 2\sqrt{\pi}\rho_0^2\frac{E_0^s(k)}{A_0(k)}\cdot \exp(2jk\rho_0). \tag{2}$$

Next, the object to be measured is placed inside the imaging area Ω, and the probe is scanned in the azimuth direction $0 \le \phi_0 \le 2\pi$ with sufficiently fine steps $\delta\phi_0$. The frequency ensures a sufficient bandwidth B and is scanned with a fine step δf and $E^s(k,\phi_0)$ is obtained. Since this scanning path is in the neighborhood of the object under test, $E^s(k,\phi_0)$ is not used directly in Eq. (1), but is the input data for NFFFT process described later. The polarization depends on the planes of the Tx and Rx antennas: vertical (V) and horizontal (H), so a total of four polarization combinations VV/VH/HV/HH can be defined. The 2D circular scanning assumes that the relationship between the Tx and Rx antennas in terms of height direction is in the far-field region. On the other hand, 1D scanning in only 1 direction is not sufficient to encompass the object and may only evaluate half-space patterns.

The basic steps of NFFFT to obtain RCS from radar images, with attention to the focusing correction process, are as follows: (i) express a scattering near-field that produces an image $\psi(x,y)$ of the object under test as a distribution of scattering centers in the image area, (ii) once the image is obtained, integrate the contributions from these scattering centers to obtain the scattering field outside the object under test, and (iii) finally, use Eq. (1) to obtain RCS in the far region.

In the following, we continue with the formulation using an electrically small scatterer as a model. This model physically gives the scattering center, and the general object can be expressed as the sum of these original scattering centers. Therefore, we believe that NFFFT described in this chapter is also applicable to general objects.

Let us derive the focusing function for 2D scanning. The radar image $\psi(x,y)$ depends on the geometry and material of the scatterer, but is independent of the measurement system such as the probe antenna characteristics. Then, we consider that the image of an electrically small scatterer placed at $(x = x_1, y = y_1)$ can be represented by the delta function

$$\psi(x,y) = C\delta(x - x_1)\delta(y - y_1), \tag{3}$$

where C is the normalized coefficient of scattering wave. From the basic fact that the scattering coefficient of all electrically small objects is proportional to the square of the frequency, the relation $A_0(k) = \tilde{A}k^2$ is obtained, where $\tilde{A}$ is mainly determined by the material composition and is independent of the wavenumber k. In the rotationally symmetric case, it is also independent of ϕ_0 (e.g., a conducting sphere of radius a is $\tilde{A} = 3\sqrt{\pi}a^3$). Now, let the normalization factor be $C = \tilde{A}$. For a fundamental scatterer with small electric length and coefficient of $A_0(k) = \tilde{A}k^2$ located at coordinate $\mathbf{i}_{\rho_0}\rho_0 = \mathbf{i}_x\rho_0 \cos\phi_0 + \mathbf{i}_y\rho_0 \sin\phi_0$, the received wave when observed at point $\mathbf{i}_\rho\rho = \mathbf{i}_x\rho\cos\phi + \mathbf{i}_y\rho\sin\phi$ on the circumference is given by

$$E^s(k, \phi_0, \rho_0) = Q(k)P^2(\beta)\frac{k^2\tilde{A}(k)}{\sqrt{4\pi}}\frac{\exp(-2jk|\mathbf{i}_\rho\rho - \mathbf{i}_\rho\rho_0|)}{|\mathbf{i}_\rho\rho - \mathbf{i}_{\rho_0}\rho_0|^2}. \tag{4}$$

The symbol $\mathbf{i}$ denotes the unit direction vector. The sum of the coherent near-fields over all azimuthal directions and frequencies is described as follows:

$$\psi(x, y) = \int_0^\infty \int_0^{2\pi} E^s(k, \phi_0, \phi)F(\mathbf{i}_\rho\rho, \mathbf{i}_{\rho_0}\rho_0, k)d\phi_0 dk, \tag{5}$$

where the weight function F, which we will understand later as the focusing operator, is given by

$$F(\mathbf{i}_\rho\rho, \mathbf{i}_\rho\rho_0, k) = g(\mathbf{i}_\rho\rho, \mathbf{i}_\rho\rho_0, k)\frac{|\mathbf{i}_\rho\rho - \mathbf{i}_\rho\rho_0|^2}{Q(k)P^2(\beta)} \cdot \exp(2jk|\mathbf{i}_\rho\rho - \mathbf{i}_{\rho_0}\rho_0|). \tag{6}$$

This function ensures that the phase and amplitude between the scattered fields received at various positions on the scanning plane are properly corrected. When the specific $g(\mathbf{i}_\rho\rho, \mathbf{i}_{\rho_0}\rho_0, k)$ is obtained later, it is found that the function F behaves as a focus correction factor in the near region.

For an electrically small scatterer placed at coordinate $\mathbf{i}_{\rho_1}\rho_1 = \mathbf{i}_x x_1 + \mathbf{i}_y y_1$, Eq. (4) is modified to

$$E_1^s(k, \phi_0, \rho_0) = Q(k)P^2(\beta_1)\frac{k^2\tilde{A}(k)}{\sqrt{4\pi}}\frac{\exp(-2jk|\mathbf{i}_{\rho_1}\rho_1 - \mathbf{i}_{\rho_0}\rho_0|)}{|\mathbf{i}_{\rho_1}\rho_1 - \mathbf{i}_{\rho_0}\rho_0|^2}. \tag{7}$$

The image function is then given by

$$\psi(x, y) = \frac{\tilde{A}}{\sqrt{4\pi}}\int_0^\infty \int_0^{2\pi} g(\mathbf{i}_\rho\rho, \mathbf{i}_{\rho_0}\rho_0, k)\frac{P^2(\beta_1)}{P^2(\beta)}\frac{|\mathbf{i}_\rho\rho - \mathbf{i}_{\rho_0}\rho_0|^2 \exp(2jks)}{|\mathbf{i}_{\rho_1}\rho_1 - \mathbf{i}_{\rho_0}\rho_0|^2}k^2 d\phi_0 dk, \tag{8}$$

where $s = |\mathbf{i}_\rho\rho - \mathbf{i}_{\rho_0}\rho_0| - |\mathbf{i}_{\rho_1}\rho_1 - \mathbf{i}_{\rho_0}\rho_0|$. Near the scatterer, $\mathbf{i}_\rho\rho \to \mathbf{i}_{\rho_1}\rho_1$, so that $\beta \to \beta_1$ and then

$$s \approx (x_1 - x)\cos\alpha + (y_1 - y)\sin\alpha \tag{9}$$

is obtained. The α refers to the angle formed by the vector $\mathbf{i}_{\rho_0}\rho_0 - \mathbf{i}_\rho\rho$ and $\mathbf{i}_x$ (see **Figure 1**), which depends on $\alpha = \alpha(\mathbf{i}_\rho\rho, \mathbf{i}_\rho\rho_0)$. Since the scanning plane encompasses the image-generating plane ($\rho_0 > \rho_{\max}$), the α increases by 2π as ϕ_0 varies from 0 to 2π. From the above, the image function is obtained by

$$\psi(x,y) \approx \frac{\tilde{A}}{\sqrt{4\pi}} \int_0^{\infty} \int_0^{2\pi} g(\mathbf{i}_\rho\rho, \mathbf{i}_{\rho_0}\rho_0, k) \exp\{2jk[(x_1 - x)\cos\alpha + (y_1 - y)\sin\alpha]\} k^2 d\phi_0 dk. \quad (10)$$

Variable transformation of the wavenumber k and ϕ_0 into $k_x = 2k\cos\alpha$ and $k_y = 2k\sin\alpha$ yields the following expression in the wavenumber domain:

$$dk_x dk_y = 4k\alpha'(\mathbf{i}_\rho\rho, \mathbf{i}_{\rho_0}\rho_0) d\phi_0 dk, \quad \alpha'(\mathbf{i}_\rho\rho, \mathbf{i}_{\rho_0}\rho_0) = \partial\alpha(\mathbf{i}_\rho\rho, \mathbf{i}_{\rho_0}\rho_0)/\partial\phi_0. \quad (11)$$

Then, the image function Eq. (10) is modified to

$$\psi(x,y) \approx \frac{\tilde{A}}{4\sqrt{\pi}} \int_{-\infty}^{\infty} \int_{-\infty}^{\infty} \frac{g(\mathbf{i}_\rho\rho, \mathbf{i}_{\rho_0}\rho_0, k)}{\alpha'(\mathbf{i}_\rho\rho, \mathbf{i}_{\rho_0}\rho_0)} \cdot \exp\{jk_x(x_1 - x) + jk_y(y_1 - y)\} k dk_x dk_y, \quad (12)$$

where the function g is defined as

$$g(\mathbf{i}_\rho\rho, \mathbf{i}_{\rho_0}\rho_0, k) = \frac{\alpha'(\mathbf{i}_\rho\rho, \mathbf{i}_{\rho_0}\rho_0)}{k\pi^{3/2}}. \quad (13)$$

This verification can be calculated as follows. Using the definition formula for the delta function [16]:

$$\delta(x - x_1)\delta(y - y_1) = \frac{1}{(2\pi)^2} \int_{-\infty}^{\infty} \int_{-\infty}^{\infty} \exp\{jk_x(x_1 - x) + jk_y(y_1 - y)\} dk_x dk_y, \quad (14)$$

and calculating Eq. (12) at $\mathbf{i}_\rho\rho \to \mathbf{i}_{\rho_1}\rho_1$ with $C = \tilde{A}$ leads easily to $\psi(x,y) = C\delta(x - x_1)\delta(y - y_1)$ in Eq. (3).

An explicit expression for the angle α can be derived by using ϕ_0. Performing the differential operation on ϕ_0, we obtain

$$\alpha'(\mathbf{i}_\rho\rho, \mathbf{i}_{\rho_0}\rho_0) = \frac{\rho_0\chi(\mathbf{i}_\rho\rho, \mathbf{i}_{\rho_0}\rho_0)}{|\mathbf{i}_\rho\rho - \mathbf{i}_{\rho_0}\rho_0|}, \quad \chi(\mathbf{i}_\rho\rho, \mathbf{i}_{\rho_0}\rho_0) = 1 - \frac{\rho}{\rho_0}\cos(\phi - \phi_0). \quad (15)$$

From this, correction factor following equation is easily expressed as

$$g(\mathbf{i}_\rho\rho, \mathbf{i}_{\rho_0}\rho_0, k) = \frac{2\rho_0^2\chi(\mathbf{i}_\rho\rho, \mathbf{i}_{\rho_0}\rho_0)}{\pi^{3/2}k|\mathbf{i}_\rho\rho - \mathbf{i}_{\rho_0}\rho_0|}. \quad (16)$$

Therefore, the focusing function is given by

$$F(\mathbf{i}_\rho\rho, \mathbf{i}_\rho\rho_0, k) = \frac{2\rho_0^2\chi(\mathbf{i}_\rho\rho, \mathbf{i}_{\rho_0}\rho_0)}{\pi^{3/2}kQ(k)P^2(\beta)} \cdot \exp\{2jk|\mathbf{i}_\rho\rho - \mathbf{i}_{\rho_0}\rho_0|\} \quad (17)$$

and the angle β of the directional antenna can be calculated as follows:

$$\cos\beta = \frac{\rho_0\chi(\mathbf{i}_\rho\rho, \mathbf{i}_{\rho_0}\rho_0)}{|\mathbf{i}_\rho\rho - \mathbf{i}_{\rho_0}\rho_0|}. \quad (18)$$

The conventional focusing function corresponding to Eq. (17) is $\chi(\mathbf{i}_\rho \rho, \mathbf{i}_{\rho_0}\rho_0) = 1$ [2, 8]. As can be seen in Eq. (15), this shows either $\rho_0 \to \infty$ (the probe is far away) or $\rho \to 0$ (the scatterer is at the origin of the coordinates), which indicates either that the probe is far away or that the scatterer is at the origin of the coordinates. The use of Eq. (16) implies that the problem can be solved for measurements in smaller anechoic chambers, that is, where the scattering center is eccentric to the origin.

The above results allow us to implicitly represent RCS. Since all radar images are distributions of point-like scattering centers, the scattered field can be expressed as a superposition of spherical waves from these scattering centers.

This can be described by the equation:

$$E^s(k, \phi_0, \rho_0) = \frac{k^2 Q(k)}{\sqrt{4\pi}} \iint_\Omega \psi(x,y) P^2(\beta) \frac{\exp\{-2jk|\mathbf{i}_\rho \rho - \mathbf{i}_{\rho_0}\rho_0)|\}}{|\mathbf{i}_\rho \rho - \mathbf{i}_{\rho_0}\rho_0)|} dxdy. \tag{19}$$

Then, paying attention to the integral transform pair in Eqs. (5) and (19), since Eq. (19) is valid for arbitrary ρ_0, the radar cross-sectional definition formula: $\sigma = 4\pi |rE^s(\mathbf{r})/E_0^i|^2$ gives the following expression in 2D by

$$\sigma(k, \phi_0) = \lim_{\rho_0 \to \infty} 4\pi\rho_0^2 \left| \frac{E^s(k, \phi_0, \rho_0)}{E_0^i} \right|^2. \tag{20}$$

Furthermore, substituting $\rho_0 \to \infty$ for $E^s(k, \phi_0, \rho_0)$ in Eq. (19) and using the relation $E_0^i = Q(k)e^{-jk\rho_0}/\rho_0$ for the incident wave, we can calculate $P(\beta) \to 1$, $|\mathbf{i}_\rho \rho - \mathbf{i}_{\rho_0}\rho_0| \approx \rho_0 - \rho\cos(\phi - \phi_0)$, which leads to the following equation:

$$\sigma(k, \phi_0) = k^4 \left| \iint_\Omega \psi(x,y) \cdot \exp\{2jk(x\cos\phi_0 + y\sin\phi_0)\} dxdy \right|^2. \tag{21}$$

This is the final formula that gives RCS at a far distance, by using the radar image in circular scanning.

If we keep the observation point ρ_0 finite and perform the NFFFT process, we can evaluate the RCS at that point. The RCS at a short distance may vary greatly with the distance. The case of anti-collision radar for automobiles corresponds to the case where the distance from the radar target is finite. For the precise design of such a radar, the evaluation of the short-range RCS of the target object provides useful information.

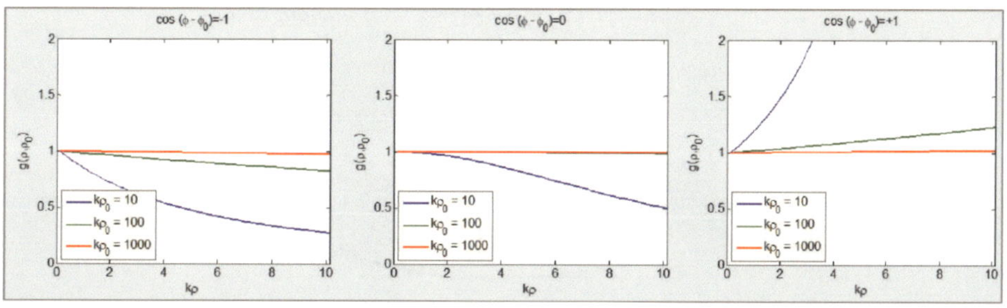

Figure 2.
Focus correcting function g for circular scanning.

As mentioned above, the improvement effect of the correction function g is expected when the scan radius is approximately equal to the target size. In other words, a large effect can be expected when the scattering center of the target is very different from the geometric center. An example of this numerical calculation is shown in **Figure 2**. In the case of $\rho \to 0$ and $\rho_0 \to \infty$, it is found to asymptote to $\rho \to \rho_0$ and $g\left(\vec{\rho}, \vec{\rho}_0\right) \to 1$ in 2D.

3. Extension from circular to cylindrical scanning

Cylindrical scanning is effective when the electrical length in the height direction of the measured object perpendicular to the circular scanning plane is large. **Figure 3** shows the coordinate system of cylindrical scanning. Basically, the same variables as for circular scanning are employed, and the z-axis is added to the circular coordinates. The cylindrical scanning NFFFT described here assumes the near-field data for monostatic RCS: $\sigma(k, \phi)$ in the horizontal plane.

As a natural extension of the basic Eq. (21) for RCS and image functions, consider the following equation:

$$\sigma(k, \phi) = k^4 \left| \int\int\int_V \Psi(x, y, z) \cdot \exp\{2jk(x\cos\phi + y\sin\phi)\} dxdydz \right|^2, \tag{22}$$

where $\Psi(\mathbf{r}), \mathbf{r} = (x, y, z)$ is the 3D radar image of the measured object, and from the scattering near-field $E^s_{near}(k, \phi_0, z_0)$, we have

$$\Psi(\mathbf{r}) = \int_0^\infty dk \int_0^{2\pi} d\phi_0 \int_{-\infty}^\infty dz_0 E^s_{near}(k, \phi_0, z_0) F(\mathbf{r}, \mathbf{r}_0.k). \tag{23}$$

$\mathbf{r}_0 = \mathbf{i}_x\rho_0\cos\phi_0 + \mathbf{i}_y\rho_0\sin\phi_0 + \mathbf{i}_z z_0$ is the coordinates of the observation point on the scanning plane, $\mathbf{r} = \mathbf{i}_x x + \mathbf{i}_y y + \mathbf{i}_z z$ is the coordinates in the image generation area V, and

$$F(\mathbf{r}, \mathbf{r}_0, k) = \frac{A_{cal}\chi(x, y, \phi_0)}{\pi^3 E^s_{cal}(k)|\mathbf{r} - \mathbf{r}_0|} \cdot \exp\{2jk(|\mathbf{r} - \mathbf{r}_0|) - \rho_0\} \tag{24}$$

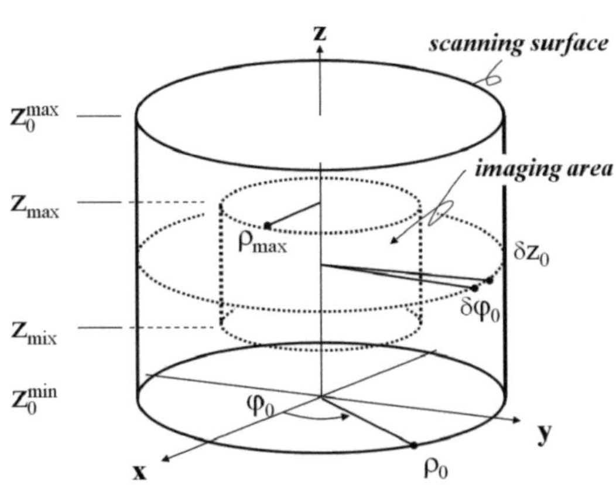

Figure 3.
Coordinate system of cylindrical scanning NFFFT: scanning surface and imaging area.

is the focusing function of the 3D cylindrical coordinate system, which, as in the case of circular scanning, can correct the differences in phase and amplitude between the scattered fields at different points on the scanning plane. In the above equation, the far-field coefficient $A_{cal}(k)$ is taken into account for the data $E^s_{cal}(k)$ from the scatterer at calibration. The function

$$\chi(x, y, \phi_0) = 1 - \frac{x}{\rho_0}\cos\phi_0 - \frac{y}{\rho_0}\sin\phi_0 \tag{25}$$

is the correction function introduced in the circular scanning. The integral expressions Eqs. (22) and (23) are calculated according to the following procedure.

First, the scattering field from the calibration scatterer is calculated as

$$E^s_{cal}(k) = E_0(k)\frac{A_{cal}(k)}{\sqrt{4\pi\rho_0^2}} \cdot \exp(-2jk\rho_0) \tag{26}$$

and the focusing function is given by the radiation field $E_0(k)$ from the aperture of the Tx antenna. Next, we consider the directivity pattern $P(\beta)$ of the probe antenna. The Tx and Rx antennas are assumed to be identical. The angle β is the angle made between the antenna boresight axis $\mathbf{i}_l$ and the direction to the point $\mathbf{r}$ on the image viewed from the point $\mathbf{r}_0$. In this case, $\mathbf{i}_l = -\mathbf{i}_{\rho_0}\rho_0, \mathbf{i}_{\rho_0}\rho_0 = \mathbf{i}_x\cos\phi_0 + \mathbf{i}_y\sin\phi_0$ can be calculated, and as a result, we have the focusing function:

$$F(\mathbf{r}, \mathbf{r}_0, k) = \frac{2\rho_0^2\chi(x, y, \phi_0)}{\pi^{5/2}E_0(k)P^2(\beta)|\mathbf{r} - \mathbf{r}_0|} \cdot \exp(2jk|\mathbf{r} - \mathbf{r}_0|). \tag{27}$$

The above equation is derived from Eq. (24) by assuming an electrically small scatterer at the point r_1. In other words, if the far-field coefficient is $A(k)$, the radar image given by the delta function $\Phi_1(\mathbf{r}) = \delta(\mathbf{r} - \mathbf{r}_1)A_1(k)/k^2$ is independent of the antenna pattern function $P(\beta)$, so the modified RCS value $\sigma_1 = |A_1(k)|^2$ is obtained.

The last one is related to Eq. (22), which can be rewritten as

$$\sigma(k, \phi) = k^4\left|\iint_\Omega \Phi(x, y) \cdot \exp\{2jk(x\cos\phi + y\sin\phi)\}dxdy\right|^2, \tag{28}$$

where Ω is the imaging region in the horizontal section of the cylindrical scan (**Figures 1** and **3**), while $\Phi(x, y)$ is defined by

$$\Phi(x, y) = \int_{-\infty}^{\infty} \Psi(\mathbf{r})dz. \tag{29}$$

Substituting Eqs. (27) and (23) into Eq. (29), we obtain

$$\Phi(\mathbf{r}) = \int_0^\infty dk\int_0^{2\pi} d\phi_0\int_{-\infty}^{\infty} dz\, E^s_{near}(k, \phi_0, z_0)\frac{\chi(x, y, \phi_0)}{E_0(k)}T, \tag{30}$$

where T of the function under integration is given by

$$T = \frac{2\rho_0^2}{\pi^{5/2}}\int_{-\infty}^{\infty}\frac{\exp\{2jk|\mathbf{r} - \mathbf{r}_0|\}}{P^2(\beta)|\mathbf{r} - \mathbf{r}_0|}dz. \tag{31}$$

Now consider the case where the probe antenna is omnidirectional, that is, $P(\beta) = 1$, so that the above equation reduces to

$$T = \frac{2j\rho_0^2}{\pi^{3/2}} H_0^{(1)}\left(2k\left|\mathbf{i}_\rho \rho - \mathbf{i}_{\rho_0}\rho_0\right|\right), \tag{32}$$

which is fully expressed using the first kind of Hankel function $H_0^{(1)}$. The vector $\mathbf{i}_\rho \rho = \mathbf{i}_x x + \mathbf{i}_y y$, $\mathbf{i}_{\rho_0}\rho_0$ are the position vector in the horizontal plane ($z = 0$). Since the Hankel function is a Green's function in free space in cylindrical coordinates, this confirms the validity of the present NFFFT theory. In addition, the horizontal position vectors other than $z = 0$, that is, the RCS pattern in that plane should be $z =$ constant. If the antenna is directional, the distance between the image generation area and the probe scanning area is $2k|\mathbf{r} - \mathbf{r}_0| \gg 1$, which is more than a few wavelengths, then it can be asymptotically evaluated. Using the stationary phase method to obtain the asymptotic solution, the result is as follows:

$$T \approx \frac{2\rho_0^2}{\pi^2 P^2(\beta_c)\sqrt{k\left|\mathbf{i}_\rho \rho - \mathbf{i}_{\rho_0}\rho_0\right|}} \cdot \exp\left(2jk\left|\mathbf{i}_\rho \rho - \mathbf{i}_{\rho_0}\rho_0\right| + j\pi/4\right), \tag{33}$$

where the angle β is given by $\beta_c = \rho_0 \chi(x, y, \phi_0)/\left|\mathbf{i}_\rho \rho - \mathbf{i}_{\rho_0}\rho_0\right|$.

Except for the measured near-field $E^s_{\text{near}}(k, \phi_0, z_0)$, all the integrand functions in Eq. (30) are independent on z_0. Therefore, the final formula for NFFFT of 3D cylindrical scanning can be given by

$$\Phi(x, y) = \int_0^\infty dk \int_0^{2\pi} d\phi_0 \left\langle E^s_{\text{near}}(k, \phi_0)\right\rangle \tilde{F}(k, \phi_0, x, y), \tag{34}$$

where the focusing function is defined as

$$\tilde{F}(k, \phi_0, x, y) = \frac{2\rho_0^2 \chi(x, y, \phi_0)}{\pi^2 E_0(k) P^2(\beta_c)\sqrt{k\left|\mathbf{i}_\rho \rho - \mathbf{i}_{\rho_0}\rho_0\right|}} \cdot \exp\left(2jk\left|\mathbf{i}_\rho \rho - \mathbf{i}_{\rho_0}\rho_0\right| + j\pi/4\right). \tag{35}$$

To calculate Eq. (34), we use the following near-field for the vertical z-axis:

$$\left\langle E^s_{\text{near}}(k, \phi_0)\right\rangle = \int_{-\infty}^{\infty} dz_0 E^s_{\text{near}}(k, \phi_0, z_0). \tag{36}$$

It can be seen that Eqs. (28) and (34) have the same form as the circular scanning NFFFT with radius ρ_0. The only difference is the representation of the focusing function [11].

Eqs. (28), (34), (35), and (36) obtained above are expected to significantly improve NFFFT processing time even for cylindrical scanning as there is no need to integrate with respect to z and z_0. The averaged electromagnetic field Eq. (36) is obtained by simply collecting data along the z-axis of the scan plane. The $E_0(k)$ frequency response of the calibration measurement is obtained from the measurement of a scatterer with a known far-field coefficient $A_{cal}(k)$, such as a conducting sphere or flat plate, using Eq. (26). Once $E_0(k)$ is known, the actual scattering field of the target can be measured, taking into account the required frequency bandwidth and scanning step.

4. NFFFT measurement software for cylindrical scanning

Figure 4 shows an overview of the software code, called KOSIP, developed by the authors to control the microwave vector network analyzer (VNA) and the scanning mechanism [9–11]. MATLAB is a convenient programming language because it is easy to create drawings and to interface with instrumentation. The KOSIP software code consists of four main windows: (i) measurement, (ii) calibration, (iii) simulation, and (ix) RCS evaluation.

In the measurement mode (i), parameters such as frequency are input, VNA, and the motor scanning system are controlled to actually measure the scattered field, and the acquired data is stored in the memory. The results of the calibration process in (ii) are incorporated and displayed in the unit (dB square meter: dBsm). This process is performed in RCS evaluation mode (ix). Simulation mode (iii) allows theoretical calculations of NFFFT with basic geometry scatterers such as conductor sphere, conductor plate, and 2- and 3-faced corner reflectors. The code has subroutines for theoretical calculation of the near- and far-fields of these scatterers, as well as dipole, pyramidal horn, for probc antcnna corrcction.

The NFFFT integrated processing code (KOSIP) has been developed for various functions and convenience [11]. In addition to monostatic radar mode, KOSIP supports bistatic mode, which is not mentioned in this chapter. ISAR processing can provide 3D images. As the theoretical analysis up to the previous section shows, it is basically indifferent what kind of object the microwave image is that evaluates the scattering pattern. This implies that the scattering pattern can be transformed as long as the microwave complex image is obtained by another measurement. It is an interesting topic to see what kind of patterns can be obtained from the images of buried underground objects or wall penetrations. On the other hand, as already mentioned, RCS patterns can also be calculated at electrically finite distances. This can be an

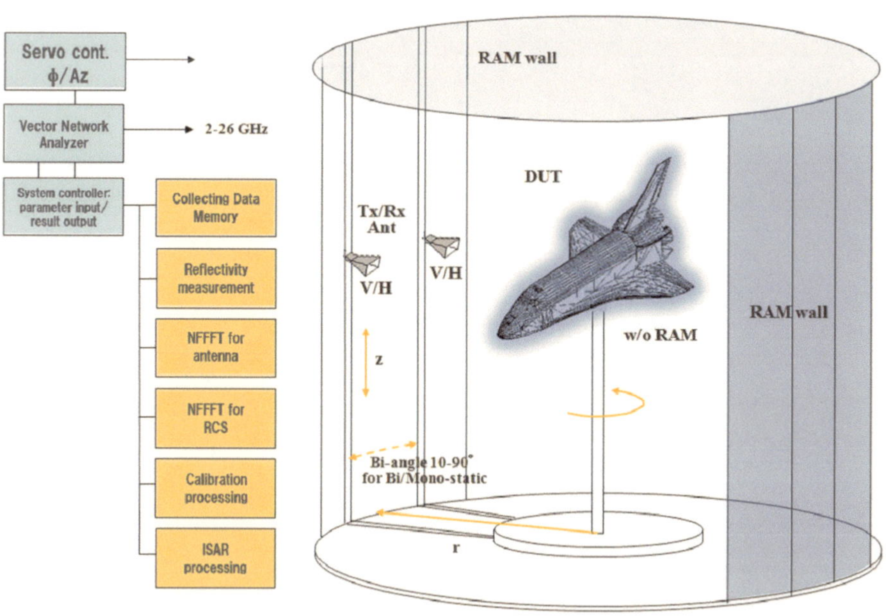

Figure 4.
Measurement system controlled by software code for cylindrical scanning NFFFT.

important point in the evaluation of RCS in short-range radar systems for collision avoidance in automobiles, etc.

As the term suggests, RCS is a measure of scattering degree from a radar with a Tx/Rx function. Therefore, RCS measurement is basically 2-way system. On the other hand, the antenna performance measurement is basically 1-way. Taking a horn antenna as an example, the far-field radiation can theoretically be obtained if the electromagnetic field distribution near the aperture plane is known. As shown in **Figure 4**, NFFFT measurement is a 2-way RCS measurement using two probe antennas. When one of the probes, for example, Tx, is connected to the feed end of the antenna to be measured, an image near the antenna is obtained. The result of the far-field transformation of this image is the far-field radiation pattern.

Examples of basic theoretical verification using the simulation mode of the KOSIP code in modeling a small multiple conducting sphere are shown in **Figures 5** and **6**. **Figure 5** shows the $|x| \leq 0.5, y = z = 0$ for 11 equally spaced small conducting spheres. The radii of the spheres are all 1.0 cm$= 1/30\lambda, f = 1$ GHz. Since the pitch between the spheres is 10 cm$= 1/3\lambda$, a row of spheres behaves like a thin flat plate (straight line). The result in **Figure 5** shows a pattern as if scattered from a flat plate, and it can be considered reasonable.

Figure 6 shows the theoretical RCS with four conducting spheres of different radii, and the center frequency is $f = 3$ GHz. The coordinates and radii of the spheres are #1(50,0,0),$r_1 = 0.1$, #2(−25,43,0), $r_2 = 0.8$, #3(−25, −43, 0), $r_3 = 0.8$, #4(−25,0,50),

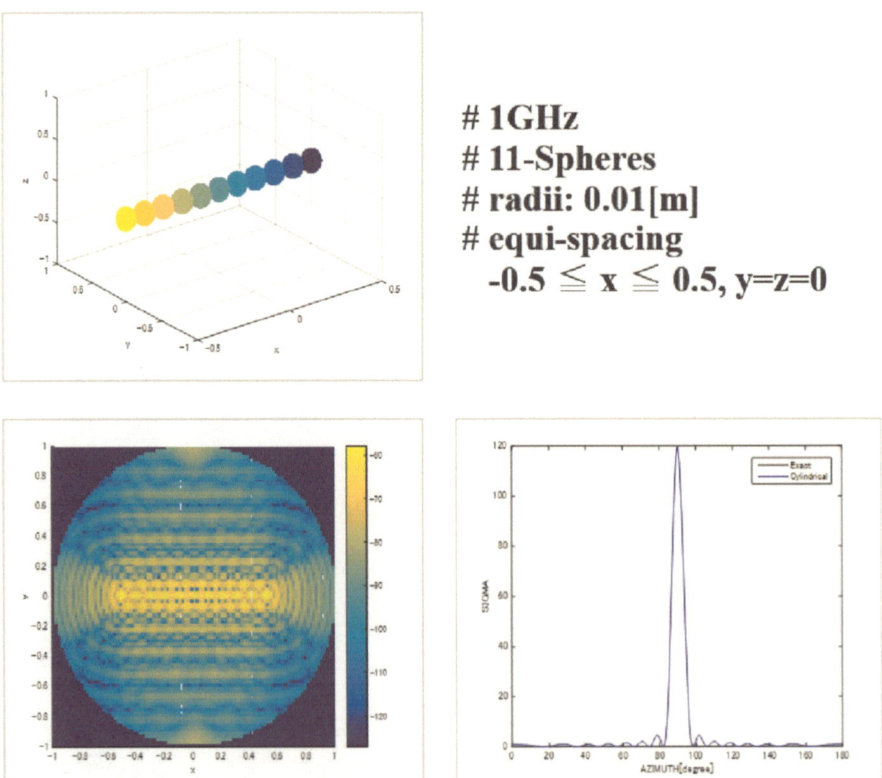

Figure 5.
KOSIP Simulation-mode: 11-spheres, radii 1/30λ, center frequency 1 GHz.

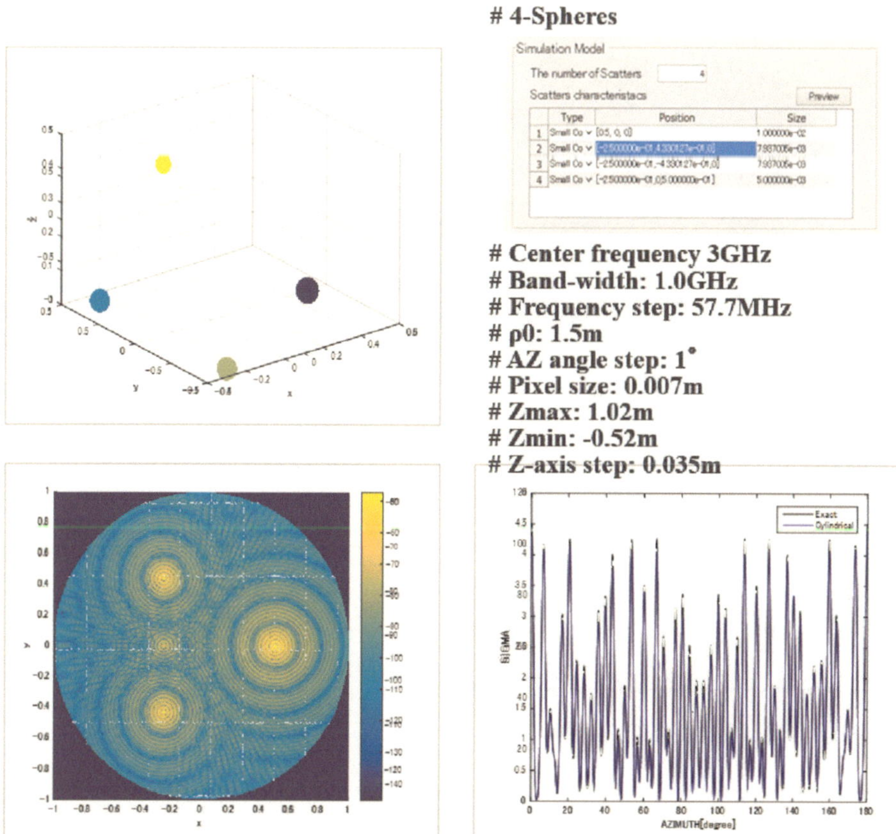

Figure 6.
KOSIP Simulation-mode: 4-spheres, center frequency 3 GHz.

$r_4 = 0.5$ in cm. The other parameters are bandwidth: 1.0 GHz, frequency step: 57.7 MHz, $\rho_0 = 150$ cm, $\delta\phi_0 = 1^\circ$, image pixel size: 0.7 cm, and $(z_{max}, z_{min}, \delta z) = (20, -52, 3.5)$ cm.

Next, we discuss the actual measurement results using RCS evaluation code NFFFT. **Figures 7** and **8** show the radar images and far-field transformed RCS measured results of a conducting sphere of 10 cm in diameter and a three-sided corner reflector with a 10 cm one side of the aperture, respectively. The measurement frequency range and step are 5.0 - 6.0 GHz and10 MHz. The scanning radius for the probe is 1.2 m, the z-axis scanning is ± 20 cm with 2.0 cm steps, and the azimuthal angular step is 1.0°. The image generation area is $\rho_{max} = 50.0$ cm, and its pixel size is 1.0 mm. In the measured example here, the direct wave removal between the Tx and Rx antennas is performed by FFT (KOSIP code is pre-loaded). It is evident that the transformed RCS pattern in **Figure 8** has the unique characteristics as a corner reflector. The measurement of the corner reflector is discussed in detail later in terms of multiple reflections.

Figure 9 shows the results of the measurements of the plastic model with conductive paint at a frequency of 5.5 GHz, a bandwidth of 1.0 GHz and a frequency step of 10 MHz. The azimuthal and vertical scanning steps are 1.0° and 2.0 cm (0.37λ), respectively, and the probe distance is $\rho_0 = 110$ cm. The image area and pixel size are $\rho_{\max} = 50\,\text{cm}$ and 0.1 cm, respectively. The two sides of the RCS pattern according to

Measurement parameter	
Target	Conducting sphere, radius 50 [mm]
Frequency	5.0 ~ 6.0 [GHz], Step 10 [MHz]
Distance	1.2 [m]
range(height)	-200~+200 [mm] , step 50 [mm]
AZ step	1 [deg]
Radar image	
Imaging region	radius: 0.5 [m]
Pixel size (x, y)	(1 [mm], 1 [mm])
Probe compensation	None
Calibration	None

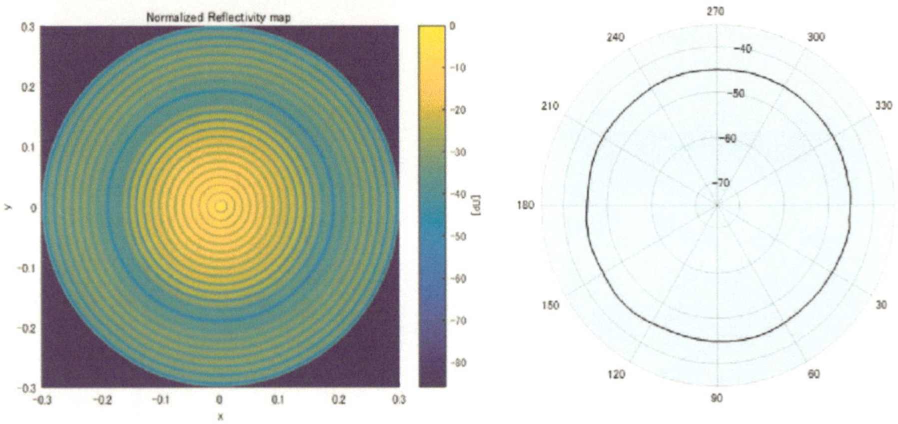

Figure 7.
Measurement of a conducting sphere by KOSIP code, cylindrical scan, center frequency 5.5 GHz.

NFFFT are symmetrical and $\phi = 0^{\circ}$ is the nose direction of the shuttle model and the maximum model length is 370 mm (6.8λ). The radar image has been improved from the initial near-field data obtained. For example, they are processed by FFT to remove the direct wave from the Tx to Rx antenna, and the image data measured without the target object are subtracted, which treats the wave as a scalar quantity and therefore requires careful handling. The former obviously includes 3D scattering data and is more valid for cylindrical scanning data than for circular. Note that the RCS pattern remains in relative dB, since calibration against certain known RCS values such as conducting spheres is omitted.

At the end of this section, some measurement considerations related to NFFFT are given as follows:

a. Spacing between Tx and Rx antennas in monostatic mode: When continuous wave (CW) is employed for Tx wave, Rx cannot receive reflected waves from the target which are smaller than the direct power from the Tx antenna. The signal isolation from Tx to Rx antenna is determined from this minimum level, which depends on the type and distance between the antennas, on the other hand, the distance is determined by $\beta_{bi} < \lambda/D_{\parallel}$ has to be guaranteed, where β_{bi} is the bistatic separation angle and $D_{\parallel}$ is the maximum horizontal target size. If the two antennas are vertically aligned, the conditional expression in this case is $\beta_{bi} < \lambda/D_{\perp}$, where $D_{\perp}$ is the maximum horizontal size of the target.

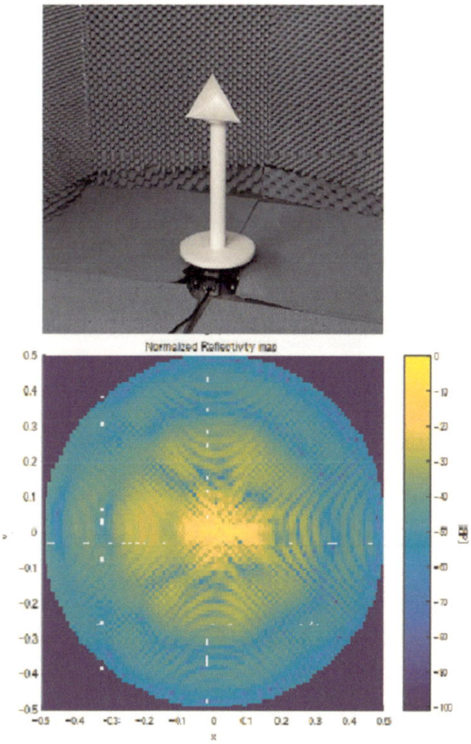

Experiment parameter	
Target	3CR(size: slant length 10 [cm]
Frequency	5.0 ~ 6.0 [GHz] (10 [MHz] Step)
Distance	1.2m
Scan range(height)	-200 ~ +200[mm] (50[mm] Step)
AZ step	1 [deg.]

ISAR parameter	
Imaging region	ρ: 1.304 [m], radius: 0.5 [m]
Pixel size (x,y)	(0.01 [m], 0.01 [m])
Probe compensation	None
Calibration	None

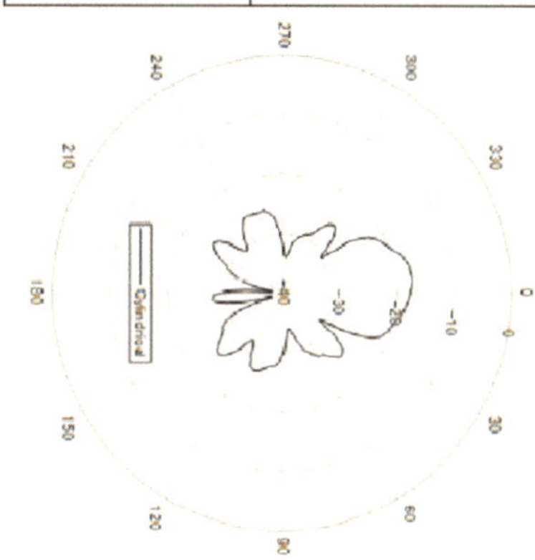

Figure 8.
Measurement of three-faced corner reflector, cylindrical scan, center frequency 5.5 GHz.

b. Level chart of measurement system: As shown in **Figure 10**, the required signal-to noise (S / N) ratio and dynamic range are very important in the real measurement environment. This depends mainly on the separation between the two antennas and the performance of the anechoic chamber. Depending on the dynamic range of the measurement system and S/N ratio, the pulsed mode measurement may be necessary when the isolation is approximately less than 30 dB. In addition, for long measurements, the VNA can be calibrated after a certain time interval to maintain the phase and amplitude at the beginning of the measurement, since the measured values may differ from the initial values.

c. Center of rotation: If the test object is a symmetrical object such as a sphere, its physical center must be as close as possible to the mechanical center of rotation.

d. Matching of NFFFT measurement and anechoic chamber: When designing a chamber quiet zone (QZ) around the target, the separate design of NFFFT microwave measurement system and the chamber environment should be avoided. Nevertheless, high-performance QZ is not considered necessary for NFFFT. A probe antenna is located near the front, so the rear wall should have low reflection.

e. Probe antenna positioning accuracy/sampling/pitch: For example, the surface accuracy of the reflector antenna should be within $\lambda/10$ rms is a desirable design condition. The reduction in antenna gain due to this tolerance is approximately

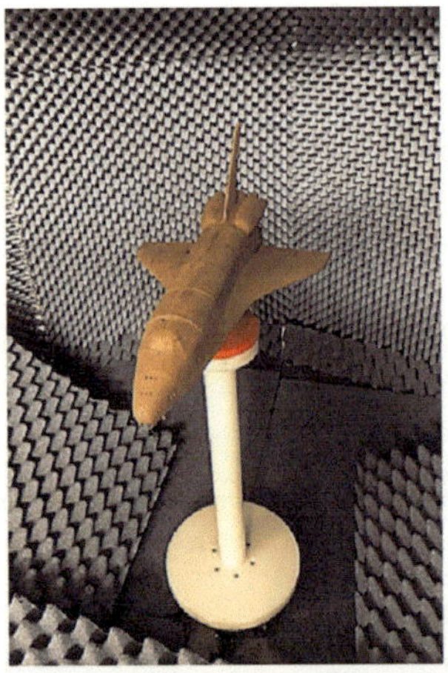

Experiment parameter	
Target	Space-Shuttle
Frequency	5.0 ~ 6.0 [GHz], Step 10 [MHz]
Distance	110 cm
Scan range(height)	-20 ~ +20 [cm] , Step 2 [cm]
AZ step	1 [deg.]

ISAR parameter	
Imaging region	Radius: 50 [cm]
Pixel size (x,y)	(0.1 [cm], 0.1 [cm])
Probe compensation	None
Calibration	None

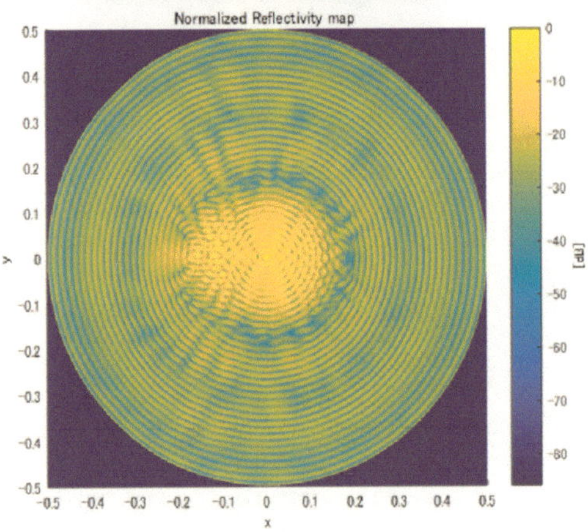

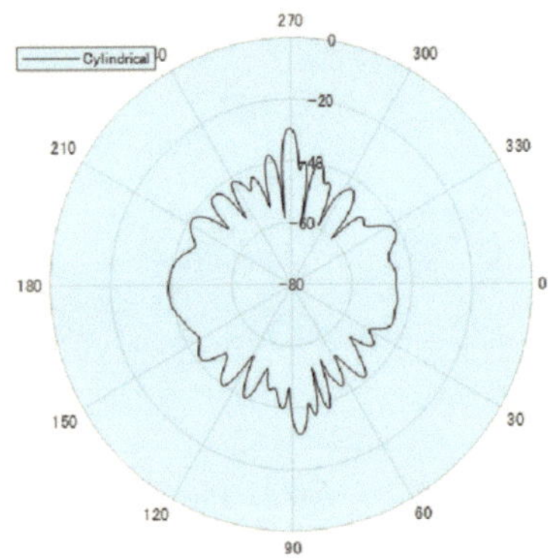

Figure 9.
Measurement of plastic-model with conducting paint, 5.5 GHz.

0.1 dB. Similarly, it is recommended that the positioning accuracy of the probe in NFFFT measurements should be kept within the same value $\lambda/10$ in both azimuth and vertical directions. On the other hand, according to the sampling theorem, all scanning sampling intervals should be kept within $\lambda/2$.

f. Multiple reflections on the target: When the target geometry is complex, such as the vertical tail of an aircraft, multiple reflections occur on the target itself. It is recommended to roughly confirm these reflection areas by ray tracing or other methods and to measure them as far away from the probing distance as possible, depending on their sizes.

g. Thinning measurement: By predicting the pattern of the electromagnetic field distribution in the z-axis, the number of sample points can be reduced toward the edges. This thinning method is very effective and contributes to reducing the time cost of measurements and calculations.

h. Fast Fourier transform (FFT): In addition to the desired scattered waves, the resulting measurement data contain all unwanted waves in the actual

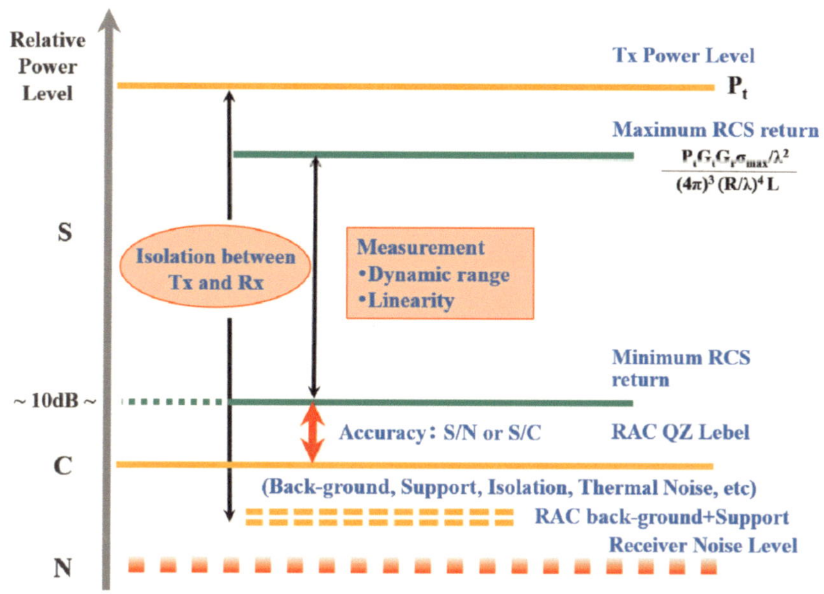

Figure 10.
Microwave-level chart: Tx power, Rx sensitivity, antennas isolation, S/N ratio, dynamic range, back wall scattering.

measurement environment. To suppress these unwanted waves, the data are usually transformed to time-domain and removed, that is, the FFT method. As described later, it has been shown that the space-domain processing of the radar image can be removed in the same way as time-domain processing by FFT.

5. Validity discussion of KOSIP-NFFFT

This section discusses NFFFT measurements and transformations for a conducting flat plate, the so-called a strip, using the geometrical theory of diffraction (GTD), the multiple reflection effect using a two-faced corner reflector, and the equivalence of the space- and time-domains.

5.1 Transformation with conducting flat plate measurement

Figure 11 shows the scattering field patterns normalized by the peak values, calculated by GTD for a conducting strip of width 22.5 cm (4.13λ). The left (a) in the figure shows the result of GTD calculation for $\rho_0 = 52$ cm (9.53λ), and (b) shows the pattern in far-field. The (c) and right (d) show NFFFT image and the far-field pattern when GTD near-field data shown in (a) is input to KOSIP, respectively.

The above theoretical results are easily confirmed by the NFFFT measurement. **Figure 12** shows (a) the photo and the parameters of the measurement, (b) the measured near-field, and (c) the far-field pattern after NFFFT. In other words, we can compare three different far-field RCS patterns with the same measurement and calculation parameters: **Figure 11(b), (d)** and **Figure 12(c)**. In order to see this clearly, we have plotted them in Cartesian coordinates in **Figure 13**. It can be seen that, except for the sidelobe region, these three patterns are remarkably consistent. There are two

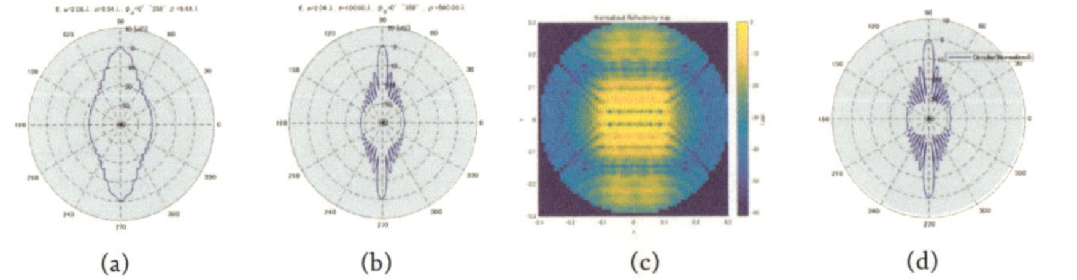

Figure 11.
GTD calculation and NFFFT processing of a conducting strip, width 22.5 cm (4.13λ), (a) GTD near-field pattern at $\rho_o = 52cm$ (9.53λ), (b) far-field pattern of (a), (c) NFFFT image of GTD near-field (a), (d) NFFFT far-field pattern of (c).

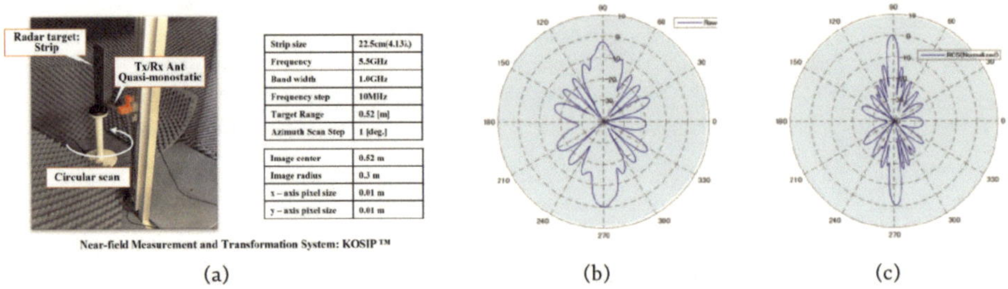

Figure 12.
(a) Near-field measurement and NFFFT processing of a conducting strip, width 22.5 cm (4.13λ) at 52 cm (9.53λ, (b) measured near-field, (c) NFFFT processing far-field pattern.

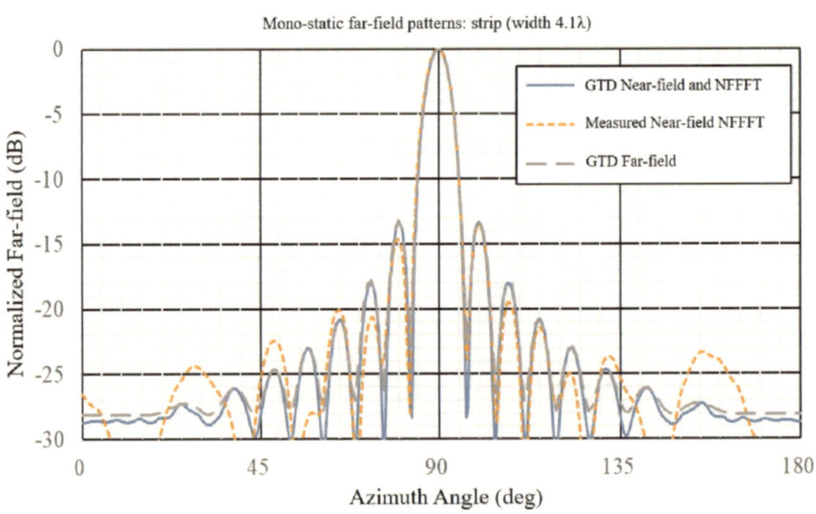

Figure 13.
*Comparison of three patterns between measurement and theory: far-field by GTD in **Figure 11** (b), GTD near-field to far-field by NFFFT in **Figure 11** (d), and measured near-field to far-field by NFFFT in **Figure 12** (c). All figures are normalized and shifted to peak angle.*

possible explanations for the discrepancy. The first is that the measurements are made in a 3D environment, while the GTD calculations are performed in 2D using a line source, and the second is that the measurements are made by circular scanning, although the length of the metal plate is finite. In any case, the results show almost perfect agreement in the main lobe and in the side lobe regions close to the main lobe. From this verification, we can conclude that the theory and processing of NFFFT is valid and effective.

5.2 Multiple reflection by two-faced corner reflector

The far-field electromagnetic field is defined by $R \geq 2D^2/\lambda$ and its corollary $D \gg \lambda$. In this section, we confirm how much error is caused by the multiple reflection NFFFT which does not satisfy the far-field condition.

Now, the width of the two-faced corner reflector used in this measurement is 15 cm on each side, and the aperture is about 21 cm, and its electrical length is approximately $D = 3.9\lambda$. Since $D \approx 4\lambda$, its far-field boundary is around $R = 2D^2/\lambda = 30\lambda = 165$ cm. Then, the distance ρ_0 between the probe antenna and the reflector aperture placed near the center of rotation is set from 50 cm (9.1λ) to 200 cm (36.4λ) in steps of 10 cm, and the near-field was obtained at each span. The distance between the Tx and Rx antennas is approximately 2λ, so that the angle β_{bi} of the object to be measured varies from 11.6° at $\rho_0 = 50$ cm to 3.1° at $\rho_0 = 200$ cm according to $\tan\beta_{bi} \approx \beta_{bi} = 2\lambda/\rho_0$. On the other hand, since $\lambda/D = 0.25$ rad$= 14.7^{\circ}$, the relation $\beta_{bi} < \lambda/D$ is satisfied for every ρ_0.

Figure 14 shows the some results of the measurements and transformations performed under the above situations. The front of the corner reflector is in the upper direction in the figure. **Figure 14(a)–(d)** shows the transformation results when the distance between the probe and the object to be measured is 50, 80, 140, and 200 cm, respectively. The left side of the each figure shows the acquired near-field (vertical polarization), the center is ISAR image from the near-field, and the right side is the normalized RCS pattern transformed from the radar image to the far-field. Although there is a difference in the width of the backward lobe in the near-field of $\rho_0 = 50$ cm, there is no significant difference between the near-field and far-field RCS at any probing distance.

Finally, we discuss the distance between the probe antenna and the target from the measurement results. The far-field distance in this measurement is $R = 30\lambda = 165$ cm. The measured distance in **Figure 14(d)** is 200 cm. In other words, the data referred to as the near boundary under these conditions are the data in the far-field, and if so, there should theoretically be no change before and after the transformation. In fact, the near boundary in the **Figure 14(d)** is almost the same as the far-field RCS. In (b) and (c), which are relatively consistent, there are some differences in lobe thickness, null shape, etc., but the validity of this NFFFT theory has been almost confirmed.

6. Equivalence of time and space to interference suppression

In relation to the multiple reflections above, unwanted reflection due to the measurement environment is discussed. In antenna or RCS measurements, one of the methods used to eliminate unwanted reflections is to first acquire the environmental characteristics without the object to be measured and then subtract them scalarly from the data when the object is placed. This method seems to work sometimes and sometimes not, if we know the vector nature of the phase. Scattering of electromagnetic waves is a phenomenon of superposition with phase change, and the reflected waves are the most prominent vector synthesis of waves with respect to incident waves. In

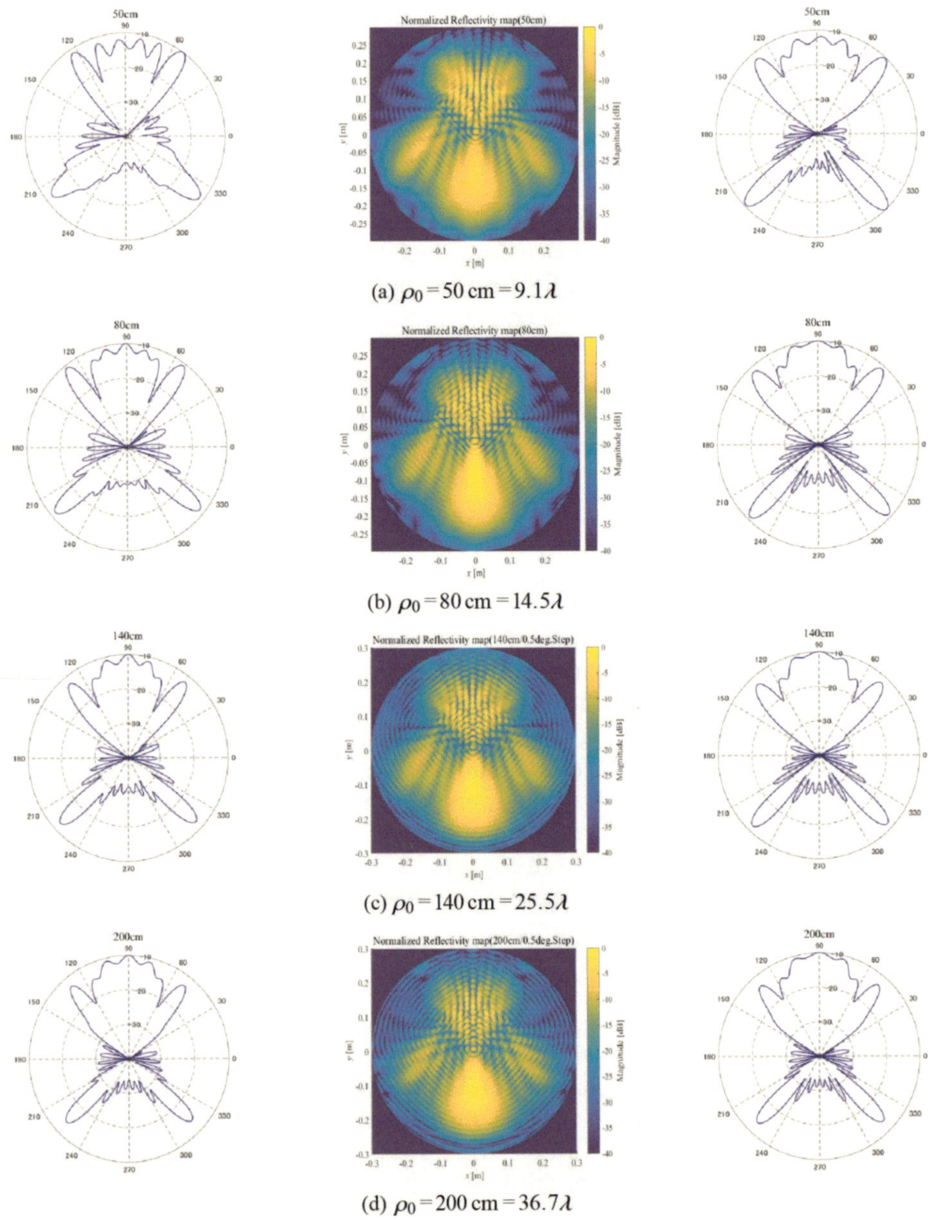

Figure 14.
NFFFT RCS measurement depending on variation of probing distance: distance of probe and target for (a)–(d) are 50, 80, 140, 200 cm, left: near-field, middle: ISAR image, right: far-field RCS by NFFFT.

this section, we discuss such multiple reflection phenomenon in the context of NFFFT measurements through simple experiments, namely the imaging region ρ_0.

Figure 15 shows the measurement phots of NFFFT processing when a small piece of metal with a size of 10×10 cm^2 is placed near the corner reflector. The distance between the probe and the center of rotation is shown as $\rho_0 = 80$ cm. **Figure 16(a)** shows the near-field acquired when the distance between the reflector edge and the small piece is 11 cm, the radar image generated (image area $\rho = 30$ cm), and the far-field RCS pattern. The image area of (a) is $\rho = 30$ cm and (b) is 20 cm. The maximum

(a) (b) (c)

Figure 15.
NFFFT measurement scenery: (a) two-faced corner reflector, (b) Tx/Rx double-ridged horn antenna, (c) small metal plate $10 \times 10\ cm^2$ *placed near corner reflector.*

distance of the small metal piece from the corner reflector is approximately two wavelengths in (a) and (b), and the effect of this multiple interference is shown around $\phi = 45 \sim 60^{\circ}$ in the front right side of **Figure 16**.

The result of the RCS transformation at the distance of 11 cm is close to **Figure 14(b)** without the additional metal plate. The only difference between **Figure 16(a)**, **(b)** is the shape of the back lobe, which is not so large. If the image area is made too small to suppress the effect of the small pieces, the image information regenerating the frontal lobes will be missing, which may affect the far-field RCS.

From the above, suppression of multiple reflections due to unwanted objects within about one wavelength is difficult in terms of image processing, and even if it is successfully suppressed (the correct answer is not known), it cannot be guaranteed at all times. In addition, the suppression of an object about two wavelengths away from the target object may spoil the desired original result if it erodes the image information

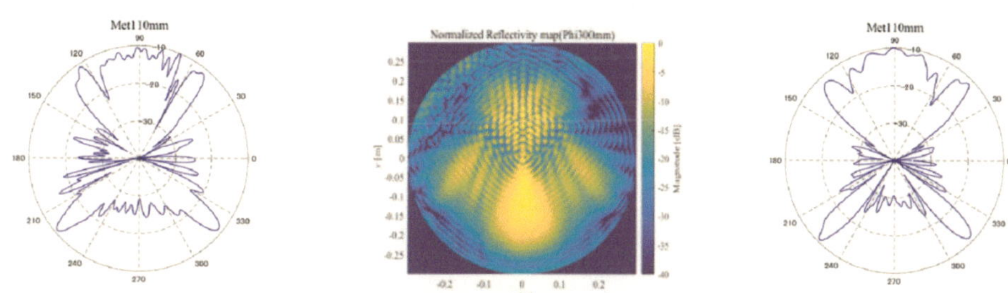

(a) distance to small plate 11 cm, left: near-field, middle: image (ρ = 30 cm), right: far-field RCS

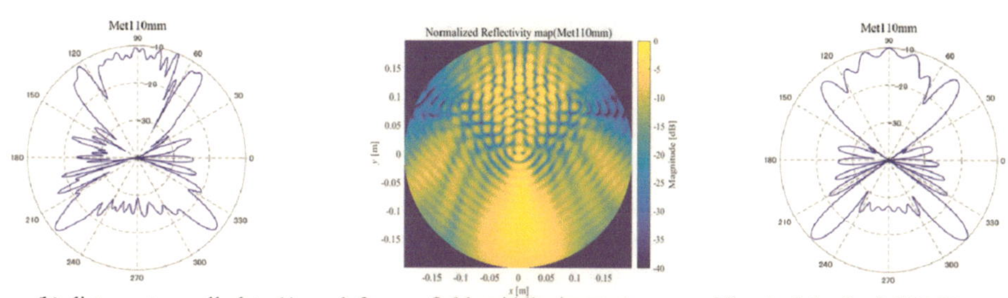

(b) distance to small plate 11 cm, left: near-field, middle: image (ρ = 20 cm), right: far-field RCS

Figure 16.
NFFFT processing of radar images of two-faced corner reflector with a small metal plate. The metal plate size: $10 \times 10\ cm^2$ *shown in* ***Figure 15*** *near the corner reflector placed at a distance of 11 cm, probe distance:* $\rho_0 = 80$ *cm, (a): imaging area* $\rho = 30$ *cm, (b):* $\rho = 20$ *cm. Near-field data of (b) is identical to (a).*

from the original object under test. It is possible to suppress the image processing when the interference level with the object to be measured is relatively small, and this can be estimated to be a few wavelengths or more away from the object. Based on these experimental considerations, there is a possibility that the scalar difference of high frequency data depending on the presence or absence of the target to be measured may result in erroneous data. Particular attention should be paid to the turntable placed near the target to be measured. Although this differencing process is generally considered to be effective in reducing the influence of the back wall of an anechoic chamber, it is not necessary in NFFFT because it is outside of the image generation area in most cases. In the extreme case, a high-precision anechoic chamber is not necessary for NFFFT. The only point to consider is whether the measurement environment is such that unnecessary high-level reflected waves are avoided.

The electromagnetic wave is a function of time and coordinates. The experiments conducted here confirm the equivalence of the electromagnetic wave in time and space. The received wave information includes all scattered waves, and it is a well-known post-processing method to remove unwanted waves by using time as a variable after transforming the wave information into the time-domain using FFT or other methods. However, this method is mainly applicable to the early and late stages of time-domain data, and is often inappropriate for processing near the desired data. On the other hand, arbitrary regions in the image space can be processed, allowing for more complex processing.

7. Conclusions

This chapter has proposed a practical RCS transformation theory for the circular/cylindrical NFFFT based on radar image information, taking into account the measurement environment, etc., through a more rigorous analysis of conventional methods, and verifies the theory by actual measurements. This theory is condensed into KOSIP code for comprehensive control and processing of the measurements. The radar image is considered as an electromagnetic field scattered by a set of electrically small scatterers, and the existence and validity of the focusing function have been clarified by analyzing the relationship between the image and the far field generated under near-field conditions.

Another well-known far-field transformation method is the compact range method. However, the bistatic measurement cannot be used for large objects such as aircraft because it requires very large and heavy reflectors and foundations. In addition, a reflector surface accuracy of less than $\lambda/10$ rms is required, and it would be impractical to manufacture very large antennas.

This chapter has also discussed the effect of multiple reflections, which is said to be the only weak point of NFFFT, based on measured data. Taking an aircraft as an example, the multiple reflections are considered as multiple reflections at each component part rather than at the entire fuselage. In the case of a two-faced corner reflector as a scatterer, the measured results show the influence of the angle of incidence on a single flat plate rather than that of multiple reflections. The suppression of multiple reflections by image processing is also considered. Although no significant effect of multiple reflections was observed in the present experimental results, further theoretical support by applying GTD and other techniques will be necessary.

Acknowledgements

The author would like to thank his collaborator Dr. Andrey Osipov (DLR, Germany) for the theoretical advice.

Author details

Hirokazu Kobayashi
Electromagnetic Wave System Laboratories, Tokyo, Japan, Former Professor at Osaka Institute of Technology, Osaka, Japan

*Address all correspondence to: h.kobayashi@kobayashi-jim.com; hirokazu.kobayashi.japan@gmail.com

References

[1] Ruck GT, Barrick DE, Stuart WD, Krichbaum CK. Radar Cross Section Handbook. NY: Plenum; 1970

[2] Broquetas A, Palau J, Jofre L, Cardama A. Spherical wave near-field imaging and radar cross section measurement. IEEE Transactions on Antennas and Propagation. 1998;**46**: 730-735p

[3] Mensa DL. High Resolution Radar Imaging. London: Artech House; 1981

[4] Osipov A, Kobayashi H, Suzuki H. An improved image-based circular near-field-to-far-field transformation. IEEE Transactions on Antennas and Propagation. 2013;**61**:989-993

[5] Kobayashi H, Osipov A. An improved image-based near-field-to-far-field transformation for cylindrical scanning surfaces. In: 30th URSI General Assembly and Scientific Symposium. 2011. pp. B04-B03

[6] Osipov A, Kobayashi H. An improved cylindrical NFFFT for compact measurement facilities. In: Proceeding of Asia-Pacific Microwave Conference (APMC2012). 2012. pp. 4D3-4D406

[7] LaHaie IJ. Overview of an image-based technique for predicting far-field radar cross section from near-field measurements. IEEE Antennas and Propagation Magazine. 2003;**45**:159-169

[8] Vaupel T, Eibert TF. Comparison and application of near-field ISAR imaging techniques for far-field radar cross section determination. IEEE Transactions on Antennas and Propagation. 2006;**54**:144-151p

[9] Kobayashi H, Osipov A, Chu C. Near-Field to Far-Field Transformation-Theory and Application. IEICE Technical Report; SANE2017–108; 2018. pp. 129-134

[10] Kobayashi H, Kameda Y, Nakamura J, Chu C. On Effect of Multiple Reflection in Near-Field to Far-Field Transformation. IEICE Technical Report; SANE2018–90; 2019. pp. 1-6

[11] Kobayashi H. RCS Evaluation by Electromagnetic Near-Field to Far-Field Transformation. Vol. 121, 16. IEICE Technical Report 2021, SANE2021–4; 2021. pp. 15-26

[12] Rudge AW, Milne K, Oliver AD, Knight P. Chapter 8. In: The Handbook of Antenna Design. UK: IEEE Press; 1986

[13] Chu TH, Lin DB. Microwave diversity imaging of perfectly conducting objects in the near-field region. IEEE Transactions on MTT. 1991; **MTT-39, no. 3**. pp. 480-487

[14] Kobayashi H, Yamaguchi Y, Cui Y. Simple near-field to far-field transformation method using antenna array factor. Journal of Wireless Networking and Communications. 2012; **2**(4):43-48

[15] Kobayashi H, Takaoka S. Near-Field to Far-Field Transformation for Radar Cross-Section and Antenna Pattern Measurement. IEICE Technical Report; AP2018–61; 2018. pp. 113-118

[16] Bladel JV. Electromagnetic Fields. 2nd ed. NY: John Wiley & Sons, Inc.; 2007

Section 3

Applications of Microwave and Electromagnetic Techniques

Chapter 5

Microwave-Assisted Thermal Conversion of Biomass: Enhancing Efficiency in Torrefaction and Gasification

Lukas Kano Mangalla and Tonni Agustiono Kurniawan

Abstract

Microwave-assisted biomass conversion, particularly in torrefaction and gasification, represents a pioneering advancement in renewable energy and waste management. This innovative technology leverages microwave radiation to achieve rapid, uniform, and selective heating, effectively addressing limitations associated with conventional thermal methods, such as prolonged processing times, energy inefficiency, and inconsistent product quality. By utilizing dielectric heating, microwave-assisted processes significantly enhance energy efficiency, improve product quality, and minimize environmental impacts. In microwave-assisted torrefaction (MAT), precise temperature control facilitates the production of biochar with superior properties, including higher energy density, improved stability, and enhanced quality. The biochar exhibits increased porosity, elevated carbon content, and greater thermal stability, making it particularly suitable for renewable fuel applications. Similarly, microwave-assisted gasification (MAG) optimizes syngas production by accelerating reaction rates, improving H_2/CO ratio, and minimizing tar formation and emissions. Comparative studies underscore the advantages of microwave-assisted processes over conventional methods, highlighting reduced tar production, lower greenhouse gas emissions, and greater scalability. By integrating MAT and MAG, microwave-assisted thermal conversion emerges as a transformative solution that aligns with global sustainability goals. This technology offers a sustainable pathway to combat climate change, reduce dependence on fossil fuels, and support a circular economy through cleaner and more efficient biomass utilization.

Keywords: waste management, microwave-assisted torrefaction, microwave-assisted gasification, energy, biomass

1. Introduction

Biomass thermal conversion is a promising technology in the global shift toward renewable energy systems. With increasing concerns about climate change, energy security, and the depletion of fossil fuels, the demand for efficient and sustainable energy generation methods has grown significantly [1, 2]. Among these methods, thermal biomass conversion—particularly through torrefaction and gasification—has emerged as a critical pathway for producing clean, renewable fuels while reducing greenhouse gas emissions [3]. However, conventional thermal conversion technologies face persistent challenges, such as low energy efficiency, prolonged processing times, and limited process control. To overcome these limitations, advanced technologies like microwave-assisted thermal conversion (MATC) have gained significant interest for their ability to enhance efficiency and product quality.

Thermal conversion processes transform raw biomass into energy-dense fuels and valuable intermediates through heat-driven chemical reactions. Among the various thermal technologies, torrefaction and gasification are the most widely explored. Torrefaction is a mild pyrolysis process conducted at temperatures ranging from 200–300°C in an oxygen-free environment, producing a solid product with enhanced energy density, hydrophobic properties, and improved grindability. This makes torrefied biomass an ideal feedstock for co-firing with coal in power plants [4–6]. In contrast, gasification involves the partial oxidation of biomass at elevated temperatures (800–1200°C) to produce synthesis gas (syngas), a versatile energy carrier that can be used for power generation, chemical production, or as a precursor for liquid fuels [7–9].

While both torrefaction and gasification offer substantial potential, they suffer from fundamental inefficiencies. Conventional torrefaction relies on slow heat transfer, resulting in uneven heating and inconsistent product quality. Similarly, gasification often struggles with biomass feedstock heterogeneity, leading to variable syngas composition and energy losses. These limitations highlight the need for innovative solutions to improve process efficiency and reliability.

Microwave-assisted thermal conversion (MATC) addresses these challenges by utilizing microwave radiation to directly heat biomass through dielectric heating. Unlike conventional heating methods, which rely on conduction and convection, microwaves penetrate the material, causing dipolar molecules—such as water—to oscillate rapidly. This internal oscillation generates uniform heat within the biomass, leading to faster, more controlled thermal processing [10]. The unique characteristics of MATC distinguish it as a transformative technology for biomass thermal conversion. Key benefits include:

1. *Rapid heating*: Microwave radiation significantly reduces processing times by enabling faster temperature increases compared to traditional methods [11].

2. *Uniform temperature distribution*: MATC ensures consistent heating throughout the material, reducing hotspots and enhancing product quality.

3. *Energy efficiency*: Directly targeting the biomass minimizes heat losses associated with external heating systems.

4. *Selective heating*: Microwave energy selectively interacts with specific biomass components, such as moisture and chemical groups, allowing for tailored thermal processes [12, 13].

In torrefaction, MATC enhances process outcomes through precise temperature control, which is essential for producing high-quality torrefied biomass with consistent energy density and hydrophobic properties. Studies have demonstrated that microwave-assisted torrefaction not only lowers energy consumption but also improves the grindability and combustion performance of the resulting biochar [14, 15]. Additionally, MATC has proven effective in processing biomass with high moisture content—an issue that typically hinders conventional torrefaction.

In gasification, MATC enables efficient and uniform heating of biomass, leading to improved breakdown of complex organic structures into syngas. Recent research has shown that microwave-assisted gasification produces higher syngas yields, with improved hydrogen-to-carbon monoxide (H_2/CO) ratios, making the process more suitable for downstream applications, such as liquid fuel synthesis [9, 16]. Moreover, MATC reduces the carbon footprint of thermal conversion processes by enhancing energy efficiency and minimizing greenhouse gas emissions, aligning with global sustainability goals and circular economy principles [17, 18].

Despite its numerous advantages, the widespread adoption of MATC faces several challenges. The high capital costs of microwave systems, including generators and reactors, remain a significant barrier to industrial-scale implementation. Additionally, biomass feedstock heterogeneity—such as variations in moisture content, particle size, and dielectric properties—can influence MATC efficiency. For instance, biomass with low dielectric properties may require pretreatment to enhance microwave absorption. Technical issues, such as the formation of "hotspots" and energy losses due to microwave reflection or transmission, must also be addressed through improved reactor designs and process optimization [19]. Furthermore, the successful integration of MATC into existing industrial systems requires careful consideration of scalability, operational compatibility, and economic feasibility.

This chapter provides a comprehensive analysis of microwave-assisted thermal conversion (MATC) in the context of biomass torrefaction and gasification. By exploring the fundamental principles, key advantages, challenges, and recent advancements in MATC, this work aims to deliver valuable insights for researchers, practitioners, and policymakers. Ultimately, MATC represents a sustainable and efficient technology for renewable energy generation and resource recovery, contributing to global efforts to mitigate climate change and transition toward a circular economy.

2. Fundamentals of microwave-assisted thermal conversion

Microwave-assisted torrefaction (MAT) offers a range of benefits that position it as a promising technology for biomass processing. One key advantage is its ability to enhance process efficiency by providing rapid and uniform heating, significantly reducing the torrefaction time while minimizing energy losses. This results in improved throughput compared to conventional methods. Additionally, biomass processed through MAT demonstrates superior quality, including higher energy density, lower moisture content, and enhanced hydrophobicity, making it highly suitable for applications such as combustion, gasification, and pelletization.

The versatility of MAT is another major benefit, as it can process a wide range of feedstocks, including agricultural residues, forestry waste, and industrial byproducts. It can also effectively handle biomass with high moisture content, reducing the need for extensive pretreatment. Moreover, the volatile compounds released during MAT can be captured and utilized as an additional energy source, further enhancing the overall energy efficiency. Lastly, MAT contributes to reduced environmental impact by optimizing energy usage, minimizing greenhouse gas emissions, and supporting sustainable biomass processing with potentially lower long-term production costs.

Despite these advantages, MAT faces several challenges that must be addressed for widespread adoption. High equipment costs, including investments in microwave generators and specialized reactors, present a significant financial barrier. Operational and maintenance expenses are also higher than those of conventional torrefaction systems. Non-uniform heating remains an issue, particularly for biomass with low dielectric properties, leading to inconsistent product quality. While microwave absorbers can alleviate this issue, they introduce additional system complexity. Feedstock variability, such as differences in density and moisture content, further impacts process efficiency and consistency, often requiring pretreatment steps that increase operational complexity. Finally, scaling MAT from laboratory to industrial applications presents challenges related to reactor design, energy input management, and system integration. Addressing these limitations is crucial to unlocking the full potential of this innovative technology.

Biomass has long been recognized as a significant renewable energy resource due to its abundance, renewability, and carbon-neutral nature. Derived from organic materials such as wood, agricultural residues, and organic waste, biomass is a versatile energy source for producing heat, power, and fuel. Currently, biomass accounts for around 15% of the global energy supply, making it an essential renewable energy source [20]. In the electricity sector, the demand for solid bioenergy is projected to reach approximately 35 exajoules (EJ) by 2050. According to the World Energy Outlook 2021, while bioenergy is expected to contribute only 5% of total electricity generation by 2050, it remains a crucial source of low-emission flexibility to complement variable generation from solar PV and wind [21]. Despite its potential, conventional thermal conversion methods, such as torrefaction and gasification, often encounter challenges such as uneven heating, low energy efficiency, and environmental concerns. These issues underscore the need for advanced technologies, such as microwave-assisted thermal conversion.

Microwave heating is an innovative technique that directly delivers electromagnetic energy to materials to generate heat. Unlike conventional heating methods, microwave heating occurs from the inside out, enabling efficient and uniform thermal processing. This approach is especially beneficial in biomass applications due to its rapid heating, energy efficiency, and ability to penetrate deeply into materials. Microwave heating operates on the principle of electromagnetic energy transfer, utilizing microwaves in the frequency range of 300 MHz to 300 GHz [22]. The most common microwave frequencies used for industrial applications are 915 MHz and 2450 MHz. The 915 MHz frequency is typically employed for large industrial applications, while 2450 MHz is more common for both industrial and domestic use. Depending on the dielectric properties of a material, microwaves may be reflected, absorbed, or transmitted. These waves interact with materials that possess dielectric properties—substances capable of absorbing electromagnetic energy and converting it into heat. When exposed to microwaves, polar molecules within the material, such as water and certain organic compounds, align with the oscillating electromagnetic field. This rapid molecular rotation generates friction, producing heat uniformly throughout the material [23, 24].

Microwave heating operates through two primary mechanisms: dielectric heating and penetration depth. Biomass exhibits dielectric properties, allowing it to absorb microwave energy and convert it into heat, resulting in the uniform heating of the bulk material. The dielectric properties of materials can be described based on the permittivity ε as follows:

$$\varepsilon = \varepsilon' - j\varepsilon'' \tag{1}$$

where ε'' is influenced by the frequency and conductivity of the material, as $\varepsilon'' = \sigma/\omega$.

Microwave can penetrate the materials, heating them uniformly compared to surface-reliant on conventional heating methods. The key dielectric properties of biomass are dielectric constant (ε'), dielectric loss factor (ε''), and the loss tangent. The dielectric constant measures the ability of material to store electromagnetic energy while the dielectric loss indicates the ability of material to convert electromagnetic energy into heat [25]. This property varies based on the type of biomass, as well as the moisture content, density, and temperature. By optimizing these factors, microwave-assisted systems can achieve precise and efficient heating tailored to the specific requirements of torrefaction or gasification processes. Loss tangent ($tan\ \delta$) indicates of how well the material can be penetrated by an electrical field and how it dissipates electrical energy as heat.

$$\tan\delta = \boldsymbol{\varepsilon}''/\boldsymbol{\varepsilon}' \tag{2}$$

The interaction between microwave energy and dielectric material can produce heat. The conversion of microwave energy into heat is known as power dissipated and can be expressed as:

$$P = 55.61x10^{-12}E^2f\boldsymbol{\varepsilon}'\tan\delta \tag{3}$$

where P is the power dissipation (W/cm^3); E is electrical field strength (V/cm); f is frequency (Hz); $\boldsymbol{\varepsilon}'$ is the dielectric constant, and $tan\ \delta$ is the loss tangent.

The governing equation of heat transfer can be modified for use in predicting heat transfer in the material placed in a microwave field. The transient heat transfer in the infinite slab can be expressed as one-dimensional formula [26]:

$$\frac{\partial^2 T}{\partial x^2} + \frac{q''}{k} = \frac{\rho C_p \partial T}{k \partial t} \tag{4}$$

where T is denotes as temperature, t is time, x is the thick of the slab, k is the thermal conductivity of material and q'' is the local heat generated by microwave.

The distribution of energy within a material is determined by the attenuation factor α' and can be calculated as follow:

$$\alpha' = \frac{2\pi}{\lambda}\left[\frac{\boldsymbol{\varepsilon}'}{2}\left(\sqrt{1 + tan^2\delta} - 1\right)\right]^{1/2} \tag{5}$$

The penetration depth of an electrical field can be calculated from the attenuation factor. The penetration depth δ_p below the surface of the material at which the electrical field strength is $1/e$ that of the electrical field in the free space, is the inverse of the attenuation factor. Thus, as it is the inverse of the attenuation factor, δ_p can be written as (where λ is the wave frequency at free space):

$$\delta_p = \frac{\lambda}{2\pi}\left[\frac{2}{\boldsymbol{\varepsilon}'\left(\sqrt{1 + tan^2\delta} - 1\right)}\right]^{1/2} \tag{6}$$

The power of microwave heating at the penetration depth can be calculated as

$$P = P_o e^{-2a'd} \tag{7}$$

Sample	ε'	ε''
Pyrolysis biochar	6.00	1.22
Oil palm char	2.83	0.23
Activated carbon	5.30	0.37
Graphite	4.20	0.11

Table 1.
Dielectric properties of several carbonaceous material at 2.45 GHz [25].

P_o is the incident power, P is power in penetration depth, d is penetration depth where $d = 1/(2a')$, and a' is the attenuation factor (**Table 1**).

2.1 Biomass composition and microwave heating

Biomass is highly suitable for microwave heating due to its complex composition, which includes water, cellulose, hemicellulose, and lignin. Water, typically present at 20–40 wt% in biomass, evaporates naturally at around 100°C, contributing to its role as a primary absorber of microwave energy due to its high dielectric constant [27].

Cellulose, a major component of lignocellulosic biomass, contains a higher carbon content than other biomass components, which significantly enhances the energy content of biomass. Its molecular structure is characterized by a strong tendency to form both inter- and intra-molecular hydrogen bonds. This results in crystalline microfibrils surrounded by amorphous cellulose, providing cellulose with superior thermal stability compared to hemicellulose [28]. Cellulose typically comprises 25–60 wt% of biomass and decomposes at temperatures ranging between 260°C and 400°C [29].

Hemicellulose, on the other hand, is more reactive than cellulose due to its branched structure, which includes functional groups like acetic acid and glucuronic acid. These branches prevent the formation of strong hydrogen bonds, making hemicellulose less thermally stable. Hemicellulose decomposition occurs at a lower temperature range, typically between 230°C and 315°C, and constitutes around 10–40 wt % of biomass [30, 31].

Lignin, another critical component of lignocellulosic biomass, serves as a natural binding agent for the structural matrix of cellulose and hemicellulose [29]. In industrial processes such as densification or pelletization, lignin acts as a natural adhesive, enhancing the mechanical integrity of the products. Lignin generally makes up 10–30 wt % of biomass and exhibits a wide decomposition temperature range of 160–900°C [31].

The dielectric properties of these biomass components play a key role in microwave-assisted heating. Water, with its high dielectric constant, absorbs microwave energy efficiently, leading to rapid heating and evaporation. Although cellulose and lignin have lower dielectric constants than water, their microwave absorption capabilities improve at elevated temperatures, facilitating the thermal conversion process [32].

2.2 Advantages of microwave-assisted thermal conversion

Microwave-assisted thermal conversion offers significant advantages over conventional biomass heating methods. Unlike traditional approaches—such as conduction, convection, or radiation—that often result in uneven temperature distribution, microwaves penetrate biomass and generate heat volumetrically. This ensures uniform heating, eliminating the occurrence of hot spots or cold zones, and leads to consistent reactions with improved product quality [33].

Energy efficiency is another key benefit, as microwaves transfer energy directly to biomass at the molecular level, minimizing heat losses associated with conventional methods. Selective heating of polar molecules further reduces unnecessary energy consumption, enhancing overall process efficiency. Additionally, microwave heating dramatically shortens processing times, enabling reactions to occur within minutes rather than hours, as seen in processes like torrefaction and gasification.

Microwaves also allow for selective heating based on the dielectric properties of biomass components. This targeted heating enables the optimization of specific processes, such as selectively heating lignin or cellulose for improved thermal conversion outcomes. The precise control over reaction conditions ensures superior product quality, such as biochar with higher energy density and hydrophobicity in torrefaction or syngas with increased hydrogen content and reduced tar formation in gasification. These attributes position microwave-assisted thermal conversion as a transformative, scalable, and sustainable technology for biomass utilization.

3. Microwave-assisted torrefaction

Torrefaction is a thermal process that involves heating biomass to moderate temperatures, typically between 200 and 300°C, under atmospheric pressure in an inert environment. The process yields three primary products: solid biochar with a dark appearance, acidic aqueous liquid, and non-condensable gaseous products. The primary objective of torrefaction is to maximize the production of biochar with enhanced properties, such as increased energy density, hydrophobicity, grindability, and storage stability [34, 35]. The resulting biochar from torrefaction has diverse applications, including briquetting, pelletization, gasification, and co-firing in thermal power plants [31].

The resulting biochar from torrefaction has diverse applications, including briquetting, pelletization, gasification, and co-firing in thermal power plants [31].

The biomass torrefaction process can be divided into five key stages, as illustrated in **Figure 1**.

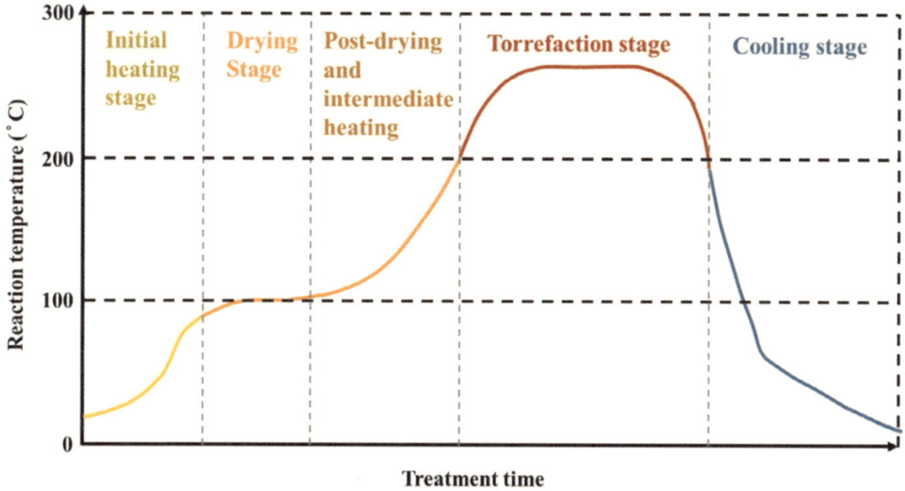

Figure 1.
Schematic stage of biomass torrefaction processes (source: Chen et al. [31]).

3.1 Torrefaction process on microwave-assisted torrefaction (MAT)

Initially, biomass undergoes heating to reach the drying stage, during which moisture gradually evaporates. At a constant temperature, free water evaporates until the critical moisture content is achieved, after which the evaporation rate slows. In the post-drying stage, further heating up to 200°C removes physically bound water and some light organic compounds, resulting in minor mass loss.

Torrefaction itself occurs at moderate temperatures and consists of three stages: heating, constant temperature, and cooling. During the heating and constant temperature phases, devolatilization occurs, leading to thermal breakdown of biomass components. The cooling phase reduces the temperature of the biomass without any additional mass loss. Torrefied biomass demonstrates several enhanced properties, including improved hydrophobicity, reduced moisture content, higher calorific value, better grindability, greater density, reduced susceptibility to degradation, and increased homogeneity [36, 37].

Conventional torrefaction relies on external heat sources and indirect heating mechanisms, which often lead to uneven thermal distribution and reduced efficiency. In contrast, *microwave-assisted torrefaction (MAT)* offers significant technological advancement by utilizing the unique properties of microwave energy to improve the process [14, 38].

In MAT, electromagnetic energy penetrates the biomass and generates heat through interactions with polar molecules. Unlike conventional heating methods, which transfer heat from the surface to the core, microwave heating occurs volumetrically. This ensures uniform temperature distribution throughout the biomass. The process is typically conducted under an inert atmosphere, such as nitrogen or carbon dioxide, to prevent combustion during torrefaction [39].

The key stages of microwave-assisted torrefaction (MAT) include the following:

1. *Feedstock preparation*: Biomass is pre-conditioned, including size reduction and optimization of its dielectric properties, to ensure efficient microwave absorption.

2. *Microwave heating*: Biomass is heated uniformly to the target torrefaction temperature. The rapid and controlled heating accelerates the decomposition of hemicellulose while partially degrading cellulose and lignin, achieving the desired thermal transformation.

3. *Biochar recovery*: The torrefied biomass is cooled in an oxygen-free environment to preserve its enhanced properties. These include increased energy density, improved hydrophobicity, and better grindability.

The integration of microwave heating and electric heating has demonstrated enhanced efficiency in torrefaction processes (**Figure 2**). A study conducted by Mangalla et al. [15] in 2024 investigated this combined heating approach using palm kernel shell as the feedstock material. In this study, microwave heating was applied for approximately 5 minutes, followed by electric heating for durations of 10, 20, and 30 minutes. The results showed a notable improvement in the energy density and grindability of the torrefied product, underscoring the effectiveness of combining these two heating methods. The schematic representation of the system is provided in **Figure 3** below.

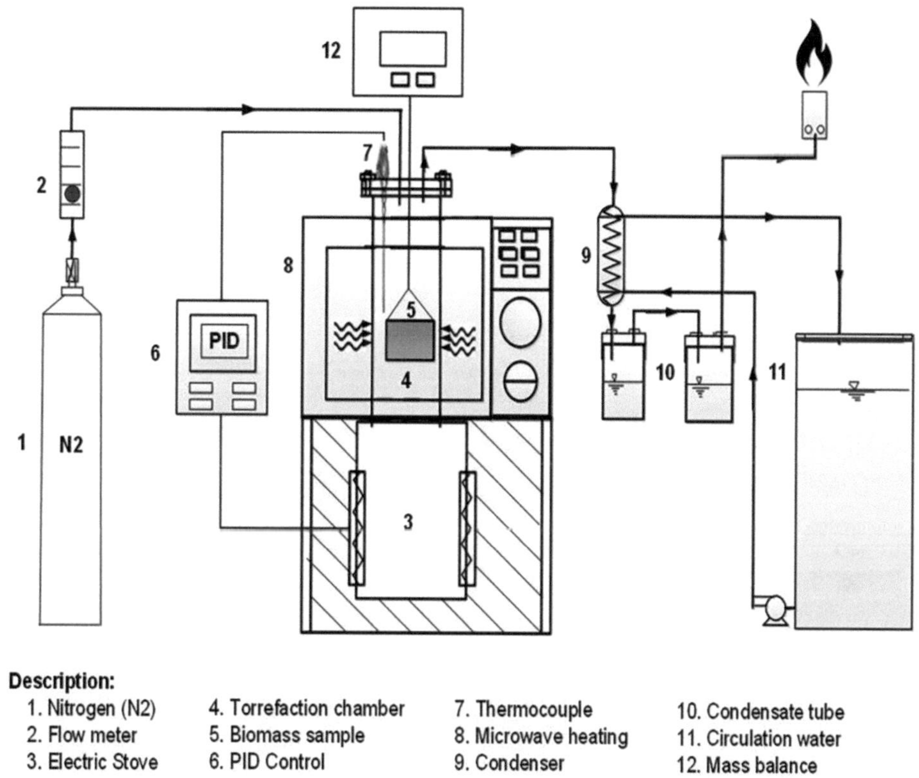

Figure 2.
Schematic research on biomass torrefaction using a hybrid electric and microwave torrefaction system (source: Mangalla et al. [15]).

3.2 Product analysis

The performance of biomass torrefaction can be assessed using parameters such as mass yield and energy yield. Mass yield indicates the proportion of the initial biomass that remains left in the solid product after the torrefaction process. Mass yield can be calculated as [40].

$$MY = \left(\frac{m_f}{m_i}\right) \times 100\% \tag{8}$$

where m_f is the mass of torrefied biomass (kg) and m_i is the initial mass of raw biomass (kg).

A higher mass yield indicates that a larger proportion of the original biomass mass is retained after torrefaction, which is generally desirable as it signifies less loss of biomass during the process. Energy yield (EY), on the other hands, refers to the amount of energy contained within the torrefied biomass compared to the raw biomass. Higher energy yield indicates better efficiency in retaining or enhancing the energy content per unit mass of biomass after torrefaction, which is crucial for evaluating the economic and environmental benefits of biomass torrefaction processes [41]. Energy yields can be written as

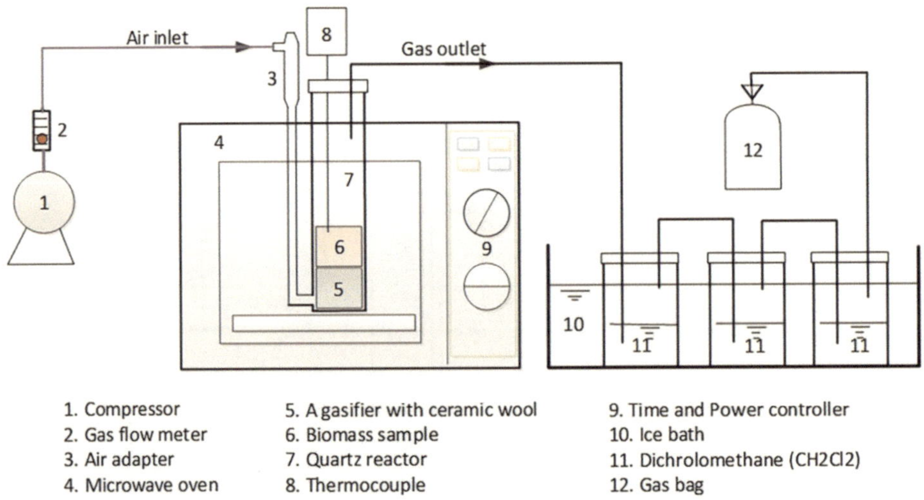

Figure 3.
Schematic view of microwave-assisted gasification (Source: Zhank, et al. [26]).

$$EY = \left(\frac{m_f}{m_i}\right) \times \frac{HHV_f}{HHV_{\mathrm{i}}} x100\% \tag{9}$$

Enhancing the energy density of torrefaction products is another crucial parameter to measures the amount of energy released when a unit mass of the torrefied product is burned. Energy density is related to terms like specific energy, calorific value, and heating value. In torrefied biomass, energy density enhancement refers to the increase in energy stored within a given volume or mass of biomass after the torrefaction process. Energy density enhancement (EDE) can be calculated using the following formula [40]:

$$EDE = \frac{HHV_f - HHV_i}{HHV_i} \times 100\% \tag{10}$$

with HHV_f represents the higher heating value of torrefied biomass (MJ/kg) and HHV_{i} represents the higher heating value of raw biomass (MJ/kg).

4. Microwave-assisted gasification (MAG)

Gasification is a thermochemical process that converts biomass or solid fuels into high-value syngas through four key stages: drying, pyrolysis, partial oxidation, and reduction. Each stage involves complex chemical reactions and physical transformations that influence the efficiency, syngas composition, and environmental impact of the process. Gasification is widely recognized as a crucial technology for biomass-to-energy applications [39]. This process transforms biomass into synthesis gas (syngas) —a mixture primarily composed of hydrogen (H_2), carbon monoxide (CO), methane (CH_4), and other gases—by reacting feedstock at high temperatures (700–1200°C) in the presence of a controlled amount of oxygen, air, or steam [9, 42].

In conventional gasification, indirect heating methods are often inefficient, leading to uneven heating and the formation of undesirable byproducts, such as tar and char. Microwave-assisted gasification (MAG) addresses these challenges through direct volumetric heating, offering a more efficient and sustainable biomass-to-syngas conversion method [26].

Gasification involves four main reaction steps, each playing a critical role in biomass conversion [43]:

1. *Drying*: Moisture is removed from the biomass feedstock through heating, typically at temperatures below 200°C. This step prevents energy losses and improves thermal efficiency in downstream stages.

2. *Pyrolysis*: Occurring between 200 and 500°C in the absence of oxygen, this step decomposes biomass into volatile gases, tar, and char. Controlling the heating rate is essential for optimizing tar formation and ensuring high-quality syngas.

3. *Partial combustion*: Operating at 700–1000°C, this stage supplies the heat needed for endothermic reactions in subsequent steps. Controlling oxygen input is critical to avoid full combustion, which would otherwise reduce syngas yield.

4. *Reduction/gasification*: At high temperatures (800–1200°C), char reacts with carbon dioxide (CO_2) and water vapor (H_2O) to produce syngas components, including carbon monoxide (CO) and hydrogen (H_2). Key reactions include the water-gas shift reaction ($CO + H_2O \rightarrow CO_2 + H_2$), the Boudouard reaction ($C + CO_2 \rightarrow 2CO$), and the water reaction ($C + H_2O \rightarrow CO + H_2$), while the methanation reaction ($C + 2H_2 \rightarrow CH_4$) generates small amounts of methane.

Microwave-assisted gasification (MAG) is emerging as a transformative technology in biomass thermal conversion. It uses microwave radiation to achieve uniform, volumetric heating, overcoming the inefficiencies of conventional methods. During MAG, biomass is processed to an optimal particle size, and its moisture content is adjusted to enhance microwave absorption. Catalysts, such as nickel or iron oxides, can also be introduced to boost syngas production [44]. The biomass is exposed to microwave energy under a controlled atmosphere (e.g., limited oxygen or steam), creating high-temperature conditions required for thermal decomposition.

The schematic diagram of a simple MAG system is shown in **Figure 4** below. MAG not only improves syngas quality but also enhances energy efficiency and reduces environmental impacts, making it a promising approach for sustainable biomass utilization.

The MAG process offers transformative advancements in biomass gasification, addressing key challenges while enhancing overall performance. It achieves critical outcomes such as high syngas yield through precise control over temperature and reaction conditions, enabling higher conversion efficiencies and maximizing valuable gas production. The rapid and uniform heating provided by microwave technology minimizes the formation of undesirable byproducts, such as tar and char, which are common in conventional gasification methods. Furthermore, MAG enhances energy efficiency by utilizing direct energy input, eliminating the need for external heat transfer, and significantly reducing energy losses. These combined benefits position MAG as a highly efficient and sustainable solution for biomass conversion.

Microwave heating promotes the water-gas shift reaction, leading to increased hydrogen production. Syngas generated *via* MAG typically has a higher H_2/CO ratio,

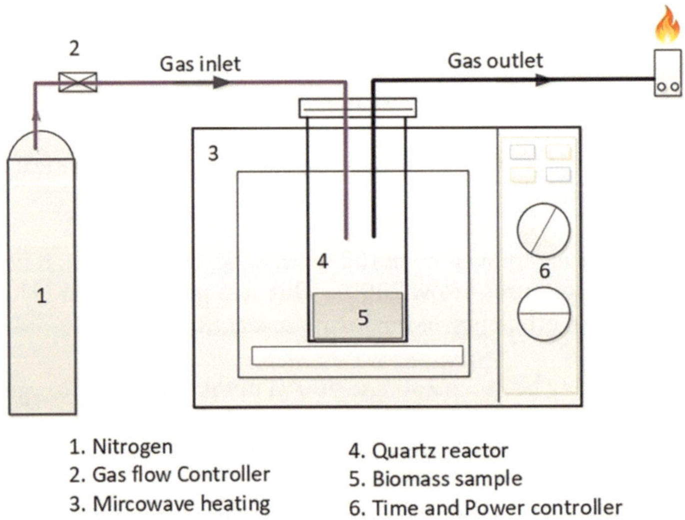

Figure 4.
Schematic view microwave torrefaction process.

which is advantageous for applications such as fuel cells or synthetic fuel production [45]. The uniform heating and high temperatures achieved in MAG ensure complete breakdown of carbonaceous materials, maximizing CO yield while minimizing unconverted char. Microwave activation of catalysts, such as nickel or potassium, enhances their efficiency, improving overall reaction rates and product selectivity [46].

Tar formation, a persistent issue in conventional gasification, is effectively mitigated by MAG. Tar, a complex mixture of high-molecular-weight hydrocarbons, can clog equipment, reduce syngas quality, and necessitate costly downstream cleanup. MAG addresses these challenges through several mechanisms [9]. First, the rapid heating rates significantly reduce the time biomass spends in the pyrolysis phase—a stage commonly associated with tar production. By quickly transitioning biomass to higher temperatures, MAG minimizes tar-forming conditions. Second, the uniform heating eliminates cold spots that often lead to incomplete reactions and tar buildup. Third, MAG facilitates catalytic reforming processes, where microwave-activated catalysts promote tar cracking, converting it into valuable syngas components such as CO and H_2. Finally, MAG's ability to maintain consistently high temperatures ensures thermal decomposition of heavier hydrocarbons, producing cleaner syngas with minimal post-processing requirements.

5. Environmental and practical advantages

Traditional biomass conversion methods rely on fossil-fuel-derived heat sources, resulting in substantial CO_2 emissions and other pollutants [47]. In contrast, microwave-assisted biomass conversion systems offer a sustainable alternative by utilizing electromagnetic energy, which can be generated from renewable sources such as solar, wind, or hydropower. This reduces carbon emissions and aligns with global decarbonization efforts. By minimizing reliance on fossil fuels, microwave-assisted systems play a crucial role in addressing climate change and improving air quality.

Microwave-assisted gasification also reduces the formation of harmful byproducts, including tar and particulates, lowering pollutant emissions and enhancing process sustainability [39]. The controlled environments, such as inert atmospheres, in microwave processes prevent the unintended release of volatile organic compounds (VOCs) and other hazardous pollutants. This enhances operational safety and ensures compliance with stringent environmental regulations, while also reducing workplace exposure to hazardous substances.

Energy efficiency is another hallmark of microwave-assisted systems. Unlike conventional methods that rely on external heat transfer, microwave technology provides direct volumetric heating. The energy is absorbed directly by the biomass, minimizing losses and ensuring a larger proportion of input energy contributes to conversion. This results in faster and more precise temperature control, enabling significantly reduced reaction times. For instance, processes like torrefaction or gasification that traditionally require prolonged heating can be completed more quickly with microwave systems, enhancing cost-effectiveness.

Additionally, microwave systems efficiently process biomass with high moisture content, eliminating the need for extensive pre-drying. Traditional technologies often require energy-intensive pretreatment, adding to costs and environmental impacts. Microwave-assisted systems simplify the process, conserve energy, and reduce operational complexity.

The combined benefits—reduced emissions, enhanced energy efficiency, and simplified processing—position microwave-assisted biomass conversion as a cornerstone technology for sustainability. By efficiently converting biomass, a widely available and renewable feedstock, into energy-dense products or syngas, these systems reduce dependence on non-renewable energy sources. This aligns with circular economy principles by promoting the reuse and recycling of organic waste into valuable energy resources.

Microwave-assisted systems also offer significant economic advantages. Shorter reaction times and reduced energy consumption lower operational costs. Eliminating extensive pretreatment steps further simplifies the process and enhances cost efficiency. The scalability of microwave-assisted systems—from small-scale applications to industrial-scale facilities—ensures their suitability for diverse biomass utilization projects, making them an attractive option for industries adopting sustainable practices.

6. Challenges and future directions of MATC

While microwave-assisted thermal conversion (MATC) holds significant promise, several technical challenges remain, particularly for large-scale applications. One major challenge is achieving uniform microwave heating in large biomass systems. Microwave interaction depends on the dielectric properties of the biomass, which can vary with moisture content and composition. This variability may cause uneven heating, leading to localized hotspots, incomplete reactions, or material degradation. Localized hotspots also risk thermal runaway, where excessive heating accelerates decomposition, further increasing temperatures and potentially damaging the reactor.

To address these challenges, precise control of microwave power levels is critical. Advanced control systems equipped with real-time monitoring and feedback mechanisms are being developed to dynamically adjust power output based on biomass properties. These systems optimize energy input by monitoring parameters such as temperature, moisture content, and reaction progress. Additionally, optimized reactor designs, including multi-mode cavities and rotating feed systems, can improve microwave distribution and ensure uniform heating.

Scaling microwave systems for industrial applications while maintaining energy efficiency is another challenge. Larger systems require higher power inputs, which can lead to heat losses and inefficiencies. Combining microwave heating with conventional methods, such as conductive or hot gas heating, can enhance system efficiency for large-scale operations. However, the higher capital costs of microwave-based systems—due to specialized equipment like microwave generators—remain a barrier to widespread adoption.

To fully harness MATC's potential, targeted research is needed in several areas. Optimizing the dielectric properties of diverse biomass feedstocks will improve microwave absorption efficiency, necessitating comprehensive databases and adaptive control systems. Catalytic advancements, particularly microwave-activated catalysts, can further boost reaction rates and reduce tar formation. Hybrid systems integrating microwave heating with technologies like plasma gasification offer opportunities to overcome current limitations by streamlining processes such as torrefaction, gasification, and tar reforming. Automation and digitalization, incorporating IoT and machine learning, can enhance process optimization and real-time adaptability. Finally, life cycle assessments and techno-economic analyses, supported by industrial-scale demonstrations, are essential to validate the sustainability and economic feasibility of MATC technologies.

Despite these challenges, MATC's potential to revolutionize biomass processing cannot be overstated. Addressing issues of heating uniformity, system scalability, and feedstock variability will unlock its full potential for sustainable energy production. With continued innovation and investment, MATC can overcome existing barriers and pave the way for widespread adoption in industrial applications.

7. Conclusions

Microwave-assisted thermal conversion (MATC) represents a significant advancement in biomass torrefaction and gasification, overcoming critical challenges such as energy inefficiency, non-uniform heating, and long processing times. By leveraging microwave heating mechanisms, MATC enables the production of high-quality torrefied biomass and efficient syngas generation, positioning it as a cornerstone technology for cleaner energy systems.

While challenges remain—including system scalability, feedstock variability, and capital costs—ongoing research and innovation offer pathways to address these barriers. The integration of hybrid systems, advanced catalysts, and digital technologies will further optimize MATC's performance and economic viability. Collaboration among researchers, industry stakeholders, and policymakers will be essential in driving adoption and unlocking MATC's full potential.

Ultimately, MATC offers a transformative approach to biomass utilization, delivering significant environmental, economic, and operational benefits. By aligning with global sustainability goals, MATC supports the transition to renewable energy and circular economy principles, paving the way for a more sustainable and cleaner future.

Acknowledgements

We sincerely acknowledge the valuable support and reference materials provided by our colleagues from the Mechanical Engineering Department of Halu Oleo University.

Conflict of interest

The authors declare no conflict of interest.

Appendices and nomenclature

C	carbon
CH_4	methane
CO_2	carbon dioxide
CO	carbon monoxide
E	electric field
Ej	exajoule
EDE	energy density enhancement
EY	energy yield
f	frequency
GHz	gigahertz
H_2	hydrogen
H_2O	water
HHV_f	higher heating value of product
HHV_i	higher heating value of raw material
m_f	mass of the product
m_i	mass of the raw material
MAG	microwave-assisted gasification
MATC	microwave-assisted thermal conversion
MAT	microwave-assisted torrefaction
MHz	megahertz
MY	mass yield
H_2/CO	hydrogen-to-carbon monoxide ratio
P	power
Po	incident power
VOCs	volatile organic compound
$tan\ \delta$	dissipation factor
λ	wave frequency at free space
α'	attenuation factor
ε''	dielectric loss factor
ε'	dielectric constant
ε	(permitivity)
δ_p	penetration depth

Author details

Lukas Kano Mangalla[1*] and Tonni Agustiono Kurniawan[2]

1 Halu Oleo University, Kendari, Indonesia

2 Xiamen University, Xiamen, China

*Address all correspondence to: lukas.kano@uho.ac.id

References

[1] Jafri N, Yoon LW, Wong WY, Cheah KH. Power generation from palm kernel shell biochar in a direct carbon fuel cell. SN Applied Sciences [Internet]. 2020;**2**(3):1-8. DOI: 10.1007/s42452-020-2189-2

[2] Mangalla LK, Endriatno N, Tasman M, et al. Utilization of waste transformer oil for alternative fuel: Design of combustion system model, economic feasibility, and performance test. Engineering and Applied Science Research. 2024;**51**:597-604

[3] Ibitoye SE, Mahamood RM, Jen TC, Loha C, Akinlabi ET. An overview of biomass solid fuels: Biomass sources, processing methods, and morphological and microstructural properties. Journal of Bioresources and Bioproducts [Internet]. 2023;**8**(4):333-360. DOI: 10.1016/j.jobab.2023.09.005

[4] Tumuluru JS, Ghiasi B, Soelberg NR, Sokhansanj S. Biomass torrefaction process, product properties, reactor types, and moving bed reactor design concepts. Frontiers in Energy Research. 2021;**9**(September):1-20

[5] Wang L, Barta-Rajnai E, Skreiberg KR, Czégény Z, Jakab E, et al. Effect of torrefaction on physiochemical characteristics and grindability of stem wood, stump and bark. Applied Energy [Internet]. 2018;**227**:137-148. DOI: 10.1016/j.apenergy.2017.07.024

[6] Grottola CM, Giudicianni P, Michel JB, Ragucci R. Torrefaction of woody waste for use as biofuel. Energy and Fuels. 2018;**32**(10):10266-10271

[7] Basu P. Biomass Gasification, Pyrolysis and Torrefaction: Practical Design and Theory. UK: Academic Press; 2013. pp. 1-530

[8] Prins MJ, Ptasinski KJ, Janssen FJJG. More efficient biomass gasification via torrefaction. Energy. 2006;**31**(15):3458-3470

[9] Arpia AA, Nguyen TB, Chen WH, Di DC, Ok YS. Microwave-assisted gasification of biomass for sustainable and energy-efficient biohydrogen and biosyngas production: A state-of-the-art review. Chemosphere [Internet]. 2022;**287**(P1):132014. DOI: 10.1016/j.chemosphere.2021.132014

[10] Zhu H, Li Y, Chen W, Huang J. Microwave pre-treatment and torrefaction of biomass: A combined approach. Energy Conversion and Management. 2022;**264**(1):115671

[11] Chen J, Li X, Gao L, Guo S, He F. Microwave treatment of minerals and ores: Heating behaviors, applications, and future directions. Minerals. 2024;**14**(3):219

[12] Chandrasekaran S, Ramanathan S, Basak T. Microwave food processing-a review. Food Research International. 2013;**52**:243-261

[13] Lozano Pérez AS, Lozada Castro JJ, Guerrero Fajardo CA. Application of microwave energy to biomass: A comprehensive review of microwave-assisted technologies, optimization parameters, and the strengths and weaknesses. Journal of Manufacturing And Materials Processing. 2024;**8**(3):121

[14] Wang Y, Sun Y, Liu X. Microwave-assisted torrefaction: Process optimization and biomass quality. Bioresource Technology. 2023;**354**(3):127231

[15] Mangalla LK, Sisworo RR, Pagiling L, Mansur. Enhancing the grindability and

porosity of palm Kernell shell through microwave preheating and torrefaction for sustainable energy production. International Journal of Intelligent Systems and Applications in Engineering [Internet]. 2024;**12**(4):1957-1966. Available from: https://ijisae.org/index.php/IJISAE/article/view/6514

[16] Ismail N, Ani FN. Syngas production from microwave gasification of oil palm biochars. Applied Mechanics and Materials. 2014;**695**:247-250

[17] Bhattacharya M, Basak T. A review on the susceptor assisted microwave processing of materials. Energy [Internet]. 2016;**97**:306-338. DOI: 10.1016/j.energy.2015.11.034

[18] Calcio Gaudino E, Cravotto G, Manzoli M, Tabasso S. From waste biomass to chemicals and energy: Via microwave-assisted processes. Green Chemistry. 2019;**21**(6):1202-1235

[19] Citra F, Almira Aulia G, Indiarto R, Hodizah Asyifaa A, Citra Angiputri Adiningsih F, Rahmalia AS. Conventional and advanced food-drying technology: A current review. International Journal of Scientific & Technology Research [Internet]. 2021;**10** (March):1. Available from: www.ijstr.org

[20] Popp J, Kovács S, Oláh J, Divéki Z, Balázs E. Bioeconomy: Biomass and biomass-based energy supply and demand. New Biotechnology. 2021;**60**: 76-84

[21] IEA. World Energy Outlook 2021. Vol. 15. IEA Publ [Internet]; 2021. Available from: www.iea.org/weo

[22] Sing RP, Heldman DR. Introduction to Food Engineering. Burlington, Sandiego, London: Academic Press; 2009. 864 p

[23] Mohd Fuad MAH, Hasan MF, Ani FN. Microwave torrefaction for viable fuel production: A review on theory, affecting factors, potential and challenges. Fuel [Internet]. 2019;**253** (April):512-526. DOI: 10.1016/j.fuel.2019.04.151

[24] Foong SY, Liew RK, Yang Y, Cheng YW, Yek PNY, Wan Mahari WA, et al. Valorization of biomass waste to engineered activated biochar by microwave pyrolysis: Progress, challenges, and future directions. Chemical Engineering Journal [Internet]. 2020;**389**(February):124401. DOI: 10.1016/j.cej.2020.124401

[25] Ellison C, McKeown MS, Trabelsi S, Boldor D. Dielectric properties of biomass/biochar mixtures at microwave frequencies. Energies. 2017;**10**(4):1-11

[26] Zhang Y, Ke C, Fu W, Cui Y, Rehan MA, Li B. Simulation of microwave-assisted gasification of biomass: A review. Renewable Energy [Internet]. 2020;**154**:488-496. DOI: 10.1016/j.renene.2020.03.056

[27] Fagernäs L, Brammer J, Wilén C, Lauer M, Verhoeff F. Drying of biomass for second generation synfuel production. Biomass and Bioenergy. 2010;**34**(9):1267-1277

[28] Onyenwoke C, Tabil LG, Mupondwa E, Cree D. Effect of torrefaction on the physiochemical properties of white spruce sawdust for biofuel production. Fuels. 2023;**4**:111-131

[29] Yang H, Yan R, Chen H, Lee DH, Zheng C. Characteristics of hemicellulose, cellulose and lignin pyrolysis. Fuel. 2007;**86**(12–13): 1781-1788

[30] Bach QV, Skreiberg O. Upgrading biomass fuels via wet torrefaction: A

review and comparison with dry torrefaction. Renew Sustain Energy Rev [Internet]. 2016;**54**:665-677. DOI: 10.1016/j.rser.2015.10.014

[31] Chen WH, Lin BJ, Lin YY, Chu YS, Ubando AT, Show PL, et al. Progress in biomass torrefaction: Principles, applications and challenges. Progress in Energy and Combustion Science [Internet]. 2021;**82**:100887. DOI: 10.1016/j.pecs.2020.100887

[32] Xu Z. Torrefaction of Mixed Solid Waste [PhD Dissertation]. Michigan Technological University; 2021

[33] Adam M. Understanding microwave pyrolysis of biomass materials. UK: Thesis Magister of The University of Nottingham. 2017;(March):1-230. Available from: http://eprints.nottingham.ac.uk/41301/

[34] Niu Y, Lv Y, Lei Y, Liu S, Liang Y, Wang D, et al. Biomass torrefaction: Properties, applications, challenges, and economy. Renewable and Sustainable Energy Reviews [Internet]. 2019;**115** (November 2018):109395. DOI: 10.1016/j.rser.2019.109395

[35] Kosowska-Golachowska M, Magdziarz A, Janus P, Urbaniak D. Experimental study on microwave torrefaction of lignocellulosic biomass. E3S Web of Conferences. 2020;**154**: 02008

[36] Martínez MG, Hélias E, Ratel G, Thiéry S, Melkior T. Torrefaction of woody and agricultural biomass: Influence of the presence of water vapor in the gaseous atmosphere. PRO. 2021; **9**(1):1-15

[37] Van der Stelt MJC, Gerhauser H, Kiel JHA, Ptasinski KJ. Biomass torrefaction: Technology and its prospects. Progress in Energy and Combustion Science. 2023;**93**(1): 100915

[38] Yan B, Jiao L, Li J, Zhu X, Ahmed S, Chen G. Investigation on microwave torrefaction: Parametric influence, TG-MS-FTIR analysis, and gasification performance. Energy. 2021;**220**

[39] Li J, Tao J, Yan B, Jiao L, Chen G, Hu J. Review of microwave-based treatments of biomass gasification tar. Renewable and Sustainable Energy Reviews [Internet]. 2021;**150**(July): 111510. DOI: 10.1016/j.rser.2021.111510

[40] Agu OS, Tabil LG, Mupondwa E, Emadi B, Dumonceaux T. Impact of biochar addition in microwave torrefaction of Camelina straw and Switchgrass for biofuel production. Fuels. 2022;**3**(4):588-606

[41] Romão DCF, Santana CC, Brito MR, Scapin E, Pedroza M, Rambo MCD, et al. Assessment of the economic and energetic potential of residues from the green coconut industry. Journal of the Brazilian Chemical Society. 2022;**33**(8): 938-947

[42] Rosyadi I, Suyitno S, Ilyas AX, Faishal A, Budiono A, Yusuf M. Producing hydrogen-rich syngas via microwave heating and co-gasification: A systematic review. Biofuel Research Journal. 2022;**9**(1):1573-1591

[43] Jara-Cobos L, Abril-González M, Pinos-Vélez V. Production of hydrogen from lignocellulosic biomass: A review of technologies. Catalysts. 2023;**13**(4):1-19

[44] Wu S, Wang Q, Wang W, Wang Y. Processes of metal oxides catalyst on conversion of spent coffee grounds into rich-synthesis gas by gasification. PRO. 2024;**12**(10)

[45] Chen GB, Wu FH, Fang TL, Lin HT, Chao YC. A study of co-gasification of sewage sludge and palm kernel shells. Energy [Internet]. 2021;**218**:119532. DOI: 10.1016/j.energy.2020.119532

[46] Ren J, Cao JP, Zhao XY, Yang FL, Wei XY. Recent advances in syngas production from biomass catalytic gasification: A critical review on reactors, catalysts, catalytic mechanisms and mathematical models. Renewable and Sustainable Energy Reviews [Internet]. 2019;**116**(October):109426. DOI: 10.1016/j.rser.2019.109426

[47] Jafri Y, Waldheim L, Lundgren J. Emerging Gasification Technologies for Waste & Biomass [Internet]. Sweden: IEA Bioenergy; 2020. 83 p. Available from: https://www.ieabioenergy.com/wp-content/uploads/2021/02/Emerging-Gasification-Technologies_final.pdf